THE MIND'S MIRROR

ALSO BY DANIELA RUS AND GREGORY MONE

The Heart and the Chip:
Our Bright Future with Robots

THE MIND'S MIRROR

Risk and Reward in the Age of AI

DANIELA RUS and
GREGORY MONE

W. W. NORTON & COMPANY
Independent and Employee-Owned

Printed in the United States of America
First Edition

For information about special discounts for bulk purchases, please contact W. W. Norton Special Sales at specialsales@wwnorton.com or 800-233-4830

Manufacturing by Sheridan Chelsea
Book design by Beth Steidle
Production manager: Julia Druskin

ISBN: 978-1-324-07932-3

W. W. Norton & Company, Inc.
500 Fifth Avenue, New York, N.Y. 10110
www.wwnorton.com

W. W. Norton & Company Ltd.
15 Carlisle Street, London W1D 3BS

1 2 3 4 5 6 7 8 9 0

CONTENTS

THE MIND'S MIRROR

Introduction

FOUR BILLION PEOPLE ON OUR PLANET CURRENTLY HAVE access to one of the most powerful and controversial technologies ever invented. Given a smartphone and a decent wireless connection, anyone anywhere can interact with an artificial intelligence system. You could ask an AI tool to write you a business plan or create an image—any picture you like. You could have it write you a new application or piece of software, even if you've never coded a single line. With a little extra thought, you could find ways to use one of these tools to help you at home, in work, and in many other parts of life.

The range of applications is truly extraordinary, and this is a very special moment in the history of artificial intelligence and indeed in the history of technology in general. Yet it is an undeniably confusing moment. There is so much misinformation, disinformation, and misguided marketing that it has become extremely difficult for the public to understand what we really have here in this technology we call AI. In our previous book, *The Heart and the Chip: Our Bright Future with Robots*, my coauthor and I focused primarily on how intelligent machines can help people with physical work and hobbies. Robots have been a particular focus and passion of mine since I was an undergraduate student, and today I'm the director of the Computer Science and Artificial Intelligence Laboratory (CSAIL) at the Massachusetts Institute of Technology. I dream of robots, and work to build a future in which they will help us all push beyond the physical limitations of our bodies.

Yet these robots and robotic enhancements would be capable of very little without the powerful artificial brains that guide their actions. As a result, AI has always been an equally critical part of my research work. I focus on both the bodies and the brains of intelligent machines, and as CSAIL director, I interact daily, and often late into the night, with many of the brightest minds in AI research around the world. These two books are in a sense companions, with one focused largely on physical work and embodied intelligence—or AI operating inside a machine that can take action in the world—and the second volume, which you're reading now, centered entirely on the artificial brains themselves and their power to help people with cognitive work. This is also a more urgent book. The robot-enhanced future we detailed has a longer timeline. The AI revolution is happening now, and it is already far more impactful than many of us expected.

My hope with this book is to not only push back against misinformation and disinformation, but to help you calibrate your expectations. These technologies are going to affect all of us, at all levels of society and business, and I believe we all need to understand them at a deeper level. *The Mind's Mirror* will explain what these varied solutions we call AI can and cannot do, and provide you with the basic knowledge and understanding needed as we all work together to shape and steer the evolution of these technologies in the years and decades ahead. The first part focuses on the different ways AI can help us as individuals and organizations—specifically, how it can allow us to work faster, access knowledge more efficiently, uncover new insights, support our creative endeavors, predict the future, master new skills, and even be more empathetic. You might think of this as the fun section. The second part details how today's predominant AI systems work, in a basic way, and offers a step-by-step plan for organizations that need to evaluate AI solutions. (This is also a fun section for geeks like my coauthor and me.) The third and final section of the book considers the unanswered and controversial questions sparked by recent developments in AI—including whether these systems will steal your job, upend society, or take over the world—and how we might go about shaping and steering AI so that it benefits the greatest possible number of people as well as our planet.

Although I do not claim to have all the answers, or perfect foresight, mine is a cautiously optimistic vision of the future. The fears and concerns are largely justified, especially with regard to economic disruption. Yet this is not like the Covid pandemic. We can see these changes coming, and if we do this right we could be facing some very exciting, even thrilling years ahead.

PART ONE

Powers

ARTIFICIAL INTELLIGENCE (AI) IS A TRANSFORMAtive force reshaping our understanding of what's possible. This technology transcends human capabilities, elevating our decision-making to new heights of efficiency and precision. AI grants us unparalleled speed, enabling us to process and analyze data at a staggering velocity. It unlocks vast reservoirs of knowledge, as it sifts through vast information repositories to bring forth insights that were once obscured in the shadows of complexity. AI enhances our creative capacities, not just by offering novel solutions but by inspiring us to generate new and unusual expressions of the human experience. With its help we can foresee various possible futures, turning uncertainty into a set of informed opportunities. As we look to master new skills and crafts, AI can serve as a companion, guiding us through learning curves with precision and patience. Moreover, AI is evolving to understand and emulate human emotions, bringing a touch of empathy to digital interactions and even teaching people how to strengthen their connections with others. These gifts—speed, knowledge, insight, creativity, foresight, mastery, and empathy—are more than tools and capabilities; they are the instruments of change in our relentless quest to improve ourselves and build a world that reflects our highest ideals and aspirations. Yet we must work closely with this technology to shape a future that reflects our shared dreams and ideals.

1

Speed

THERE ARE TIMES I WOULD LIKE TO BE AS FAST AS WONder Woman or the Greek goddess Nike—when I'm late for an appointment, for example, or eager to get home after a long day at work. In those instances, it would be nice to outrun a train or burst through the sound barrier. But if I were to choose speed as a superpower, what I would really like is the ability to finish my work faster. I'd like to accelerate the pace at which I brainstorm with colleagues, conduct research, and develop plans to implement new ideas. In fictional stories, we see characters achieve this kind of accelerated mental productivity through magical accidents or make-believe pills. The 2012 movie *Limitless* features a failing novelist who ingests the wonder drug NZT-48, which amplifies his brainpower and allows him to complete a masterpiece of fiction in merely four days. Yet we don't need imaginary neurological performance enhancers. We have AI.

In the real world, we are developing AI tools capable of endowing people with accelerated productive powers. These tools will enable us to finish various cognitive tasks in much less time. They will change how we work at the office, how we approach domestic chores, even how scientists search for cures. Think of AI as a multitalented, expert personal assistant enabling all of us—marketers, investors, scientists, lawyers, coders, and even parents doubling as domestic engineers—to operate faster. To be clear, I do not advocate offloading cognitive work to AI systems entirely—for example, typing out a prompt to generate a report instead of doing the research and writing yourself. Yet you could lean on an AI tool to brainstorm approaches after you've studied and developed your argument, thus accelerating your work. A shopping assistant could track and order your groceries, then suggest recipes based on what's in your fridge. At work, you could have the equivalent of an overqualified intern that sifts through tons of information, rapidly pulling out the key points you need for a project or assignment. This quick-thinking, all-knowing partner could help you make decisions by swiftly analyzing data and spotting patterns that might take a human a lot longer to see.

Or consider what I'm doing here—writing. Naturally I was tempted to see whether AI might accelerate the process of creating this book, but I had a few reasons for resisting the urge. The first is technical. The tools that have drawn so much interest in recent years, like ChatGPT and Bard, are in some ways highly advanced versions of the auto-complete function on your smartphone. They predict the likely next word while tracking higher-level structure. (There is a great deal more complexity involved—the systems represent these words as tokens in a multidimensional space, as I will explain in more detail in Part Two—but let's stick to the simple definition for now.) And they are phenomenally capable. So if I were to ask one of these language models to, say, write a book chapter about how AI is going to accelerate productivity in many areas of life, the program would be able to generate something that looks and even reads appropriately because it has studied trillions of words in countless documents. There would probably be hallucinations, or invented facts, and I might very well disagree with some of the claims. I'd have to revise heavily, but it would certainly be fast. The main problem, though, is that the book would not really represent my ideas. It would be a carefully selected string of text modeled on trillions of words spread across the web. My goal is to share *my* knowledge, expertise, passion, and fears with regard to AI, not the global average of all such ideas. And if I want to do that, I cannot rely on an AI chat assistant.

Tools like ChatGPT, Bing Chat, Bard, Claude2, DALLE, Gemini, LiquidAI, Midjourney, and others are a form of generative AI—a family of systems that can produce digital art, music, videos, and text of unexpected quality. While they cannot write my book, or help a novelist turn out a masterpiece in four days, for shorter works they can be truly valuable assistants. In one study of how generative AI impacts writing, for example, scientists recruited 444 college-educated professionals, including marketers, consultants, grant writers, managers, and others in similar roles. Each individual was given two writing tasks, such as drafting a report, an email, or a press release. They were informed that they'd be rewarded with bonus payments if they produced quality writing. The end products of their labor were judged by professionals working in those same jobs. Half the participants were told they were allowed to use the popular generative AI tool ChatGPT to complete their second task, and were trained in the basics of how to use it. Roughly four out of five chose to do so.

What happened? Well, the less experienced writers in the study benefited in several ways. When they used ChatGPT, their writing samples were of a higher quality relative to their unassisted pieces, as judged by professionals, and they finished their assignments faster. The highly skilled writers who used ChatGPT didn't produce better work, but they did produce writing of similar quality faster. The time devoted to creating a rough draft was cut in half. Plus, the participants reported that they were able to focus more on their core ideas and the revision process that sharpens the expression of those ideas. They'd done the research. They'd used their brains. The tool assisted them with exposition—it helped them write what they knew. Overall, the work produced with the help of AI was of equal or better quality, *and* it was completed faster.

Personally, I'd like a more intelligent email assistant, and I suspect that any manager, department head, or business executive would benefit from such a tool. I receive hundreds of emails each day and often spend a few hours reading and answering them in turn. What I'd like is a solution that reads, summarizes, and generates quality responses for me to review and revise. This assistant would need to rely on a foundational large language model—the kind of AI engine that powers interfaces like ChatGPT—trained on billions or trillions of words. But I don't want my assistant to sound like everyone else. I want it to sound like me and reflect my way of thinking, my style of addressing issues, and my general operating principles as the director of a prominent AI lab. All of this information is present within my emails, but we'd need an intelligent tool to extract these patterns. So I'd take the standard language model and develop an additional, Daniela-focused variation that operates atop the foundational piece. To do so, I'd gather the hundreds of thousands of emails I have stored in my system and train this new model on them in order to identify patterns that correspond to my stylistic preferences, writing voice, and decision-making principles. When I sit down at night to answer my emails, I don't want the system to answer for me and send the emails directly. I need to be responsible for every message. But if I could start with a draft to review and revise before clicking Send, I'd easily double my email response pace.

A system like this could be adjusted or tuned to the needs of doctors and nurses who are overburdened with administrative tasks. Extensive paperwork is required to ensure high-quality patient care, adhere to compliance regulations, and generally operate the larger healthcare system in an efficient manner. Yet these passionate, highly educated, and intensely trained individuals did not pursue medicine because they love paperwork. So why not tune and train systems that listen to their interactions with patients, access recorded doctor–patient interactions, vital sign measurements, and health records, and use this collected information to generate notes and fill out forms faster? The healthcare professional would still need to review and certify everything, but this could free up significant time for more direct patient interaction.

The startup company Codametrix* is using AI to accelerate one of the slowest and most expensive phases of the administrative cycle—medical coding. Any time a provider seeks reimbursement from an insurance provider, detailed forms must be filled out that include very specific codes to classify the case. There are more than 100,000 of these codes, so the task of filling out the documentation has traditionally fallen to specialists. Recently, the number of specialists began to dwindle, and the manual coding task was becoming slower and more expensive. Codametrix trained an AI solution to read a healthcare provider's notes and

* My husband is one of the cofounders.

reports, then automatically select the appropriate code and fill out the requisite forms. The technology proved to be five times faster and results in 60% fewer denials for reimbursement relative to manual coding.

In the insurance industry, one interesting startup found a way to apply similar tools to fill out forms that would normally waste the valuable time of highly paid adjusters. My students are fond of an AI insurance tool from the company Lemonade, which pays out approved claims in seconds. They love the intuitive interface, core technology, and of course the rapid payouts. The legal industry has seen widespread adoption of productivity-accelerating AI tools, but lawyers have also provided a test case for what can go wrong. In June 2023, the judge P. Kevin Castel of Federal District Court in Manhattan fined a pair of attorneys for using ChatGPT to generate a brief—or written summary of their legal argument—in support of their client's lawsuit. Unfortunately, they were a little too reliant on the technology, which included several past cases that provided precedents for their argument. When the judge read the brief, he tried to look up the cases cited as reference points, and he couldn't find them. The reason? They did not exist. ChatGPT had made them up. When questioned, the lawyers confessed that they had not checked the references themselves. Instead, they'd asked ChatGPT if the cases were real. The system responded in the affirmative.

• • •

This acceleration effect will not be limited to writing tasks or filling out forms. We're seeing similar trends in software development, as the same fundamental AI tools that accelerate the writing of stories can speed up the writing of code. These tools work a little like the autocomplete on your phone. As a programmer types out code, the AI system suggests the rest of the line. If the developer likes what she sees, she can accept the suggestion and move on, thereby working a little faster. The system also makes it easier to find and insert large chunks of mundane but necessary boilerplate code. When GitHub, the company behind the popular coding tool Copilot, surveyed its users in an early study to find out what they thought about working with the AI, they posted some startling results. Of the developers surveyed, 88% said they completed tasks faster, and 96% claimed to be faster with repetitive tasks. How much faster? GitHub recruited ninety-five developers to complete a programming task, asking half to do it manually and half to use Copilot. Those who used the AI assistant finished the job in one hour and eleven minutes on average. Those who worked entirely on their own needed two hours and forty-one minutes. The majority of the developers who used the tool said they were happier as well. I'd be happier if I worked that much faster, too.

These speed-enhancing abilities stretch beyond generative AI. Consider the field of drug discovery. Typically it takes years for researchers to develop a new drug candidate for a disease. In 2022, a group of scientists at the University of Toronto and a company called Insilico Medicine set out to see if they could accelerate this timeline with the help of AI. First they needed a target, and they chose a form of cancer that killed an estimated 782,000 people in 2018 alone. The scientists deployed an AI system to scour data from studies on the cancer, and it identified a collection of twenty proteins that appeared to play a significant role in its progression. The scientists reviewed the list of twenty and narrowed it down to a single promising target. Proteins are the workhorses in the body, for good and bad. So, by identifying a protein that looked to be particularly influential, and then developing a drug that could shut down this culprit's activity, they might have a new way to slow or even stop the cancer. Think of this approach as the equivalent of targeting and containing the best player on an opposing football team in order to win.

The scientists had to understand the protein at a deeper level to truly affect its work, so they turned to AlphaFold, a deep learning system developed by DeepMind, a subsidiary of Alphabet Inc. AlphaFold predicts protein folding, or the process by which a protein chain acquires its functional, three-dimensional structure.* You need to have a sense of the structure to understand the function of the protein within an organism. Predicting folding has been a longstanding challenge in the field, and AlphaFold is a major breakthrough with myriad applications in biology and medicine, from understanding diseases to developing new drugs. In 2021, DeepMind and the European Bioinformatics Institute partnered to release AlphaFold's predictions for many

* AlphaFold is like a sophisticated puzzle-solver that predicts the 3D shapes of proteins. Think of proteins as long chains of beads, where each bead represents an amino acid; these chains twist and fold into complex 3D structures. To understand how proteins fold, AlphaFold begins by studying a vast library of known protein shapes, much like how a child learns to recognize objects by seeing many examples. Given a new protein, AlphaFold predicts how close or far apart each pair of beads might be and what angles they might form with each other. It's akin to predicting how a string of beads will arrange itself when dropped on a table. Using these predictions, AlphaFold constructs a 3D model of the protein, like trying to re-create the shape of the beads on the table by knowing only the distances and angles between them. The initial model might not be perfect, so AlphaFold tweaks it to make sure it makes sense according to the laws of physics, adjusting the beads' positions so they don't overlap and the string isn't too stretched or compressed. Finally, AlphaFold checks the accuracy of its predicted shape by comparing it to the actual shape of the protein (if available), similar to comparing a child's drawing of an object to the real thing.

protein structures, making them freely available to the scientific community, which brings us back to the drug discovery experiment. The researchers tapped into AlphaFold's knowledge of the protein's structure to identify potential weak spots. Then they tuned another AI model to suggest designs for drugs that would latch on to those weak spots and inhibit the activity of the protein. Once again, the researchers reduced the results, narrowing the field of AI-generated drugs to seven promising candidates. Finally, they moved to the lab and actually made the molecules, synthesizing them through chemical reactions and other complex processes. After testing these products, they ultimately settled on a single compound—the one with the highest likelihood of success. This entire process and its many steps undoubtedly sounds very complex. Yet it was remarkably fast. Instead of the years normally required to identify a new compound, the discovery took merely thirty days.

My point here is not to imply that AI is poised to cure cancer. In fact, the research team behind this study did not explicitly set out to bring a drug to market. This was about potential. They were showing what may soon be possible as we apply AI to the different stages of drug development. Generally, this is also a good example of how we should think about making use of AI to accelerate processes in all areas of discovery, work, and life. The researchers did not merely lean back and sip their coffee and let the AI do everything. Their process required repeated interaction between the human scientists and their various intelligent tools. The first model suggested possible proteins, which the scientists narrowed to one. The second identified potential attack points on the protein. The scientists applied their knowledge of biochemistry to pick from these options. Finally, when the AI provided ideas for compounds that might prove to be effective drugs, the scientists used their expertise to choose which ones to synthesize and test.

So, AI did not find this potential drug. Scientists used AI to discover it faster.

• • •

AI tools will not merely accelerate how we write, work, program, or do science. They could have a subtle but important influence domestically, too. Consider dinner. When I return home after work, I want to make something delicious and healthy. I enjoy cooking, and I have many cookbooks in my home and love each one. When I have time to peruse them in advance and go to the market to shop for the best ingredients, the process is wonderful. But who has that kind of time on a weeknight? Typically, I arrive home in a rush and need to quickly figure out what to prepare based on what we have in our pantry and fridge. Searching the Internet for an appropriate recipe is frustrating; you end up with too many options, and it takes far too long to sift through them and find one that works with your ingredients, tastes, and timing. So I'd like an AI-powered cookbook instead.

How would we build that? An effective AI solution is like a house in that it needs a foundation. For our intelligent cookbook, as with my email assistant, this would be a large language model that has been trained on vast amounts of digital text. Once we have the foundational model in place, we could fine-tune it on all the recipes freely available on the web, or we could narrow the scope and create a collection based on my library of cookbooks. Either way, with the combination of the foundational language model and the corpus of recipes, we'll be on our way to chatting with our AI Cookbook.

Let's imagine you arrive home famished. First you tell the AI Cookbook what sort of food you'd like—gluten-free, for example, or vegetarian—and how much time you want to spend preparing and, if necessary, cooking the meal. Eventually, your fridge and pantry may be smart enough to track which ingredients you have at home, and in what quantities, which would allow the AI to narrow the possible recipes and only suggest options that match your existing ingredients. But your AI Cookbook could also ask you the relevant questions directly, and by informing the system that you would like to make something from a dozen fingerling potatoes, half an onion, a dash of cream, and other ingredients, your options would be narrowed and focused. Then you could inform your assistant that you'd like something light and healthy—maybe Indian?—and that you'd prefer to avoid gluten. Perhaps you'd specify that you only want to spend fifteen minutes on preparation. Instead of you going back and forth between your cookbooks and pantry, your assistant would provide you with viable options in seconds. The AI Cookbook would take in your parameters, scour its library of recipes, synthesize this information, then generate an ideal meal plan.

The synthesis and generation pieces are critical here; this is where we start to move into this new age of AI. If you were to simply search for an optimal match for your ingredients, tastes, and timing, you'd really just be using a version of the sort of information retrieval systems we've all been interacting with for decades. This would be a refined search engine with a filter on inputs (ingredients), processing (prep time), and outputs (hot vs. cold dish, Indian vs. Chinese, sweet vs savory, etc.). A truly intelligent cookbook would ingest these same requirements and generate something entirely new. This system could find hidden patterns in successful recipes and scour reviews to build a model of what sort of flavor profiles and ingredient combinations to embrace and avoid. Would the suggested recipe actually taste good? I'm not sure. We haven't built the system yet.

This is not science fiction; these applications are possible now. I'm not suggesting the computer science equivalent of wormhole travel. We really can use AI as a kind of mental accelerant and productive superpower today, and these tools are going to have an impact across many personal tasks and industries. We're already seeing a difference. In manufacturing, AI-driven design tools can be used to create on-demand tooling and new parts, increasing the speed and accuracy of production. The delivery service UPS employs an AI-based navigation system to optimize routes by factoring in weather, traffic, package volume, and other variables, thus speeding up deliveries. The logistics industry that enables so much of our online ordering and rapid shipping is changing now that AI can forecast demand, optimize inventory levels, and generally ensure that the right products get to the right place at the right time. Meanwhile, Amazon's Go stores use AI and computer vision to accelerate cashier-less shopping, allowing customers to walk right through checkout lines without even stopping to scan items. AI-powered inventory management systems can also automate restocking processes in retail stores, reducing manual labor and ensuring optimal inventory levels. Predictive algorithms can support transportation fleet management by forecasting maintenance needs and optimizing scheduling, reducing both fuel consumption and travel time. Radiologists are reading and interpreting medical images such as X-rays or MRIs faster with the help of AI algorithms, speeding up the diagnostic process and fine-tuning accuracy.

Another example is the work of my MIT colleague Hamsa Balakrishnan's group; they are developing an AI-assisted scheduling system to help the U.S. Air Force schedule crew for its C-17 military cargo aircraft. Each flight typically requires a crew of six. There are 275 aircraft in the fleet, and in 2021, the Air Force needed to assign crews to operate the planes for 4 million flight hours. This is a very complex problem, as they need to account for everything from rest requirements to vacation, and Hamsa's solution makes it faster to determine optimal schedules. The AI system, which integrates with a preexisting tool, suggests potential schedules, complete with crew and pilot assignments. The algorithm at the heart of the AI scheduler receives a numerical reward when a schedule is accepted. Over time, it picks out patterns in these successful, reward-earning schedules, and optimizes its suggestions.*

The fact that these tools accelerate productivity does not mean we should deploy them with equivalent speed. The integration of AI requires careful planning, including understanding which tasks can be solved better with AI and which ones should remain the responsibility of our very capable biological minds. There are many issues to address related to data privacy, security, ethics, bias, and workforce training. And while I understand the general concern regarding building solutions that are increasingly intelligent, it is important to remember that in doing so, we are also making ourselves smarter. The large language models that have drawn so much attention and sparked so much controversy contain much of human knowledge, and as such they might be one of the most powerful tools for accessing information that we have developed in centuries.

* This process is called reinforcement learning, which I will discuss in detail in chapter 10.

2

Knowledge

I STILL LOVE AN OLD-FASHIONED LIBRARY, WHERE INFORmation meets inspiration, and the inquisitive mind can find nourishment and the seeds of creativity and scientific discovery. The Harry Elkins Widener Library at Harvard University, with its ten levels and three and a half million volumes, is endlessly stimulating, but my favorite lies on the other side of the Atlantic. While I was attending a conference at the Royal Society in London, my colleague and friend Sir Mike Brady arranged a tour of the Society's library for several of us attendees. This library houses some of the most extraordinary handwritten gems in the history of science: Isaac Newton's manuscripts (including drafts and notes related to his groundbreaking work in mathematics, physics, and astronomy); Charles Darwin's early ruminations on his theory of evolution by natural selection; Edmond Halley's handwritten memoranda regarding his astronomical observations; the observations of Captain James Cook; letters from Albert Einstein, Michael Faraday, and other prominent scientists; and "An

Essay Towards Solving a Problem in the Doctrine of Chances" by Thomas Bayes. (This last one might not sound quite on par with Newton or Darwin, but as the founding father of probability theory, Bayes is something of an intellectual rock star for computer scientists.) The archivist even brought out a first edition of Newton's *Philosophiae Naturalis Principia Mathematica*, in which the famed scientist and mathematician introduced his version of the calculus.

This was a singular experience, yet there is one other library I've always longed to visit, and unless one of our MIT students has an unexpected breakthrough with time travel or builds its virtual digital twin, I don't expect to get the chance. I would like to explore the Great Library of Alexandria, the central repository of human knowledge in the ancient world. Naturally it would be inspiring to stroll through this institution, immersing myself among the works and scrolls crafted by some of history's greatest minds. I'd like to study a manuscript or two and engage with the scholars of the time. Yet I'd also like to visit for a very particular reason. I am very curious about how they organized all that knowledge. Neither the card catalog nor the Dewey Decimal System had been invented. How were the scrolls arranged and stored? Who was allowed to read them? And how did one go about requesting access? I'd like to imagine there was an all-knowing librarian who could tell you exactly where every single scroll was located, precisely what it contained, and perhaps even synthesize the contents for you succinctly.

Today, the digital transformation of information has given us access to a nearly infinite library. This has created a problem of abundance. The number of scientific papers alone has doubled every fifteen to seventeen years, a rate that has continued for more than a century and appears to show no signs of slowing. In robotics and AI we have gone from less than 1,500 papers released in 2015 to more than 6,500 in 2021. And that was before the AI revolution began! You don't have to be a scholar to feel overwhelmed by all this information. This is a common affliction for curious minds. There are so many new ideas and studies and articles emerging all the time that it has become almost impossible to keep up. And this is merely the world of digital media. Information is distributed across the natural world, from our immediate physical surroundings to the distant cosmos, waiting for us to extract and transform it into knowledge. The universe is in some sense a vast library, and we desperately need an AI version of my fictional Librarian of Alexandria.

Let's call it the AI Librarian.

Before I describe this tool, allow me to zoom out for a moment. The wide range of digital material spread across the Internet is a mix of data and information. Think of data as raw and unfiltered, without any meaning assigned. When this raw data is processed, organized, structured, or presented in a given context, it becomes information. A random collection of numbers would be data, but if those numbers were organized into a table representing monthly sales figures for different products, then you'd have information. The information science and knowledge management theorist Russell Ackoff believed these concepts could be organized into a hierarchical pyramid with data at the bottom, information above that, and knowledge above that. You gain or acquire knowledge when you interpret or find patterns or trends in the information—for example, when you look at those monthly sales results and see which products are performing better than others. Today we have an overabundance of data and information. What we need is a more efficient way to help us sift and transform this excess into personal knowledge.

By synthesizing information, the AI Librarian could save you the time of reading countless pages and digital tomes, accelerating your knowledge acquisition and helping you become well-versed in almost any field. Why is this different from what we have today with Internet search? A standard information-retrieval search engine fetches information for you. This approach was fine for the first decade or two of the Internet age and will continue to be satisfactory for many use cases, but it amounts to what is called an information pull. You need certain information. You enter a specific query in a search engine. Then the system goes out and finds what appear to be the appropriate pages, ranks them, and provides you with links in order of relevance. The system *pulls* the information from the Internet and makes it available to you, like a librarian retrieving a specific book for you from the stacks. What we want is more of a push—a tool that intelligently suggests and shares information when you don't know exactly what you need, or when the information becomes available rather than when you ask for it, more like a proactive and deeply experienced research librarian.

The need for such human experts will not diminish. Yet they are limited in number, and these tools will extend such services to more people. How would we build this digital librarian? There are many different variations of AI that could be useful, but with large language models, we already have an incredible resource and the opportunity to magnify our capabilities further. Think of these tools as both microscopes and telescopes. The AI microscope reveals knowledge that would be challenging to discern independently in much the same way that a biological microscope unveils details invisible to the naked eye. Similarly, as a telescope allows astronomers to view distant celestial objects, AI extends our vision into potential futures, scenarios, and possibilities that are otherwise beyond our immediate perception. In both instances, the microscope and the telescope, the user has to interpret and make sense of the information revealed.

The language models are trained on the written sum total of humanity's knowledge—in principle, everything that has ever been written by anybody, everywhere, at any time. They've swallowed up the equivalent of the Library of Alexandria, Widener, the contents of the Royal Society, and the information available on the Internet, including various sources of misinformation and crackpot theories, too. Even if the models do not truly understand anything that they read and merely use this text to uncover patterns and links between words and strings of words, the information still exists within them. And with the right tool, we can extract this information and turn it into knowledge.

The startup Perplexity is developing an impressive service that allows users to ask any question, then rapidly provides an answer with easy-to-check citations and sources. A query about when life first appeared on our planet, for example, returns a clear and concise answer drawn from the website pages of the Smithsonian Institution, the University of Chicago, and other trustworthy organizations. At MIT, Pattie Maes and David Karger have embarked on a project to create a personalized AI library inspired by Memex, an early idea for a machine capable of storing, linking, and retrieving large volumes of information. Memex was the brainchild of the legendary engineer, inventor, and early government science advisor Vannevar Bush. Back when he conceived the idea, in 1945, the technology needed to realize his vision didn't exist, but Pattie and David have conceived of a modern variation using large language models. This is an ideal application because AI models capture, process, and organize vast amounts of data, making connections between seemingly disparate pieces of information. Imagine having all your books, notes, and records in a single, searchable database that shows how different pieces of information relate to one another. To achieve this, we would need a robust data storage infrastructure, an intuitive interface (think voice commands or touchscreens, or in the case of Pattie and David's project, natural language), and technologies like natural language processing to understand human-generated content. We would also need machine learning and AI tools to automate data organization and make personalized recommendations. Ensuring privacy and security has to be a top priority, requiring strong encryption and user authentication. Yet these are solvable problems and would be well worth the effort. With a technology like this, you would no longer be held captive by memory blanks. You'd have an assistant at the ready, prepared to help you recall that story, anecdote, name, or fact you know is hiding somewhere in your brain.

The critical difference, relative to the current broadly available tools, is that this would be finely tuned to your personal knowledge, information, and experience. This is good, but also limiting. None of us know everything, not even the teenagers among us. So we do want to be able to access the vast digital informationscape, and in this sense the models that have effectively swallowed the Internet are tremendously exciting as knowledge-growth tools. One of the downfalls of these tools, however, is that they did not study only verified, scholarly work. They also swallowed misinformation, vitriol, and conspiracy theories. In some sense what we need is an AI-enhanced version of Wikipedia, the compendium of human knowledge that has been synthesized and verified by people. This AI variation would have to incentivize the people contributing to the corpus to continue working on the medium to ensure its continued veracity and reliability. For example, we could layer our AI Librarian atop Wikipedia in such a way that the general knowledge returned would be verifiable, with sources cited and easy-to-check references.

This approach, with Wikipedia specifically or digital libraries in general, would not put librarians out of work or even endanger the profession. Instead, it would empower them, as they could use these AI tools to assist students and curious patrons in their search for knowledge; librarians could use their virtual colleagues to become instant experts in any subject and play an increasingly critical gatekeeper role as well. The answers returned by our AI Librarian would have to be vetted, because the system could potentially make mistakes or invent facts. In the future, could these systems get to the point where you would simply trust what they generate without having to oversee them? Maybe. But we're not there yet.

The legal profession provides another alternative. The innovation in this space has been remarkable. In the past, an attorney researching a case might have visited the firm's law library or asked a senior partner for advice. Later, legal libraries were digitized, and lawyers could search through the vast electronic corpus of laws and cases as they built or refined an argument. The legal service Westlaw, for example, is compiled from more than 40,000 databases containing potentially relevant information including case law, statutes, public records, law journals, and other sources. As with a standard information retrieval system, interacting with such tools is more of a pull than a push. But it pulls from a quality database.

The lawyers who were fined in federal court for relying on ChatGPT ran into trouble because that highly advanced chat engine did not have the same filter on quality. ChatGPT drew from all publicly available information leading up to September

2021, much of it not verified or fact-checked. ChatGPT itself did not know the difference between fact and fiction, so it invented judicial opinions. However, though this application was misguided, AI can be a powerful research assistant for lawyers when designed correctly. The company Casetext developed a tool called CoCounsel that relies on OpenAI's GPT-4, the foundational large language model that powers ChatGPT Plus. The integration with GPT-4 makes it easy for lawyers to interact with their AI research assistant using natural language. In other words, it allows them to *talk* to the system. At Casetext, the engineers and lawyers who trained the technology spent 4,000 hours refining the model through what we call reinforcement learning with human feedback. Basically, they queried the model in early tests, evaluated the answers it returned, re-ranked the results to move the optimal ones to the top, and gave this information back to the system. In doing so, they showed the model the better answers. The model then adjusted to increase its chances of matching those results and getting it right the first time. We teach infants the same way. Think of a toddler speaking her first words. She attempts to say a given word and we then pronounce it properly in hopes she will gradually correct her enunciation.

The reason this particular solution works so well is that it only searches the verified, up-to-date legal database that Casetext has been building and refining since 2013. CoCounsel can't get creative or turn to GPT-4 for facts. The tool either cites real, specific, verified legal information or it does not return an answer. The resulting solution can now review documents, search databases for relevant cases, suggest topics for lawyers to focus on when deposing witnesses, summarize contracts or longwinded legal opinions, and much more. All these results still need to be reviewed and checked, even though the tool works with verified databases, but this highly qualified assistant spares lawyers the sort of work that swallowed up too many hours in the past and allows them to focus on higher-level reasoning. By spending less time chasing down relevant documents, they have more time to think and strategize.

These subject-focused AI librarians could be applied to many other domains, including medicine, academia, and the home. When one of our many devices or appliances stops functioning effectively, my husband pores through the instruction manual or pdf to find the solution. He genuinely enjoys reading manuals. I do not. What I'd prefer is to overlay an AI trained on both the contents of the manual and verified customer cases published on the web. Instead of moving back and forth between the index, table of contents, and videos when something goes wrong with our dishwasher or a car, I'd like to be able to simply talk to the manual, sharing details of the breakdown, then prompting it to diagnose the problem and generate potential solutions. This would be difficult to build, but very possible.

These research assistants could extend to our hobbies as well. My parents love to garden. I do as well, or I would like to in theory, but in practice I'm terrible—so bad, in fact, that I once killed a cactus. Not directly or intentionally; I neglected this incredibly robust and resilient plant, which evolved to survive in the desert, to the point at which it could not even survive in my home. Yet my parents grow all kinds of vegetables, from peas to tomatoes, and we are all too willing to partake of this bounty at family dinners.

At one point, my father decided that the soil at my home would be ideal for growing fruit trees and grapevines to make delicious liqueurs. When one of the trees appeared to be suffering from some kind of affliction, he snipped off a leaf, brought it to a gardening center, and asked one of the employees how he might go about curing it. The person he spoke with was helpful, but not an expert in the field. Yet that knowledge is out there! In 2016, for example, a trio of scientists trained a neural network on more than 54,000 images of plant leaves representing fourteen different crop species, all labeled as either healthy or diseased. The model was able to determine with 99.35% accuracy whether a plant was sick or not. Since then, developers have launched apps with such capabilities. If my father had had something similar loaded onto his smartphone, perhaps with a feature that specified which type of disease was the likely cause, he could have found his answer and begun looking for solutions immediately. He could have become the expert through the device in his pocket. This is an example of how AI can connect digital knowledge to the physical world.

When I was a teenager, my friends and I were enchanted with a series of five novels written by the Romanian author Constantin Chiriță. The books follow a group of adventurous kids who call themselves the Cherry Blossom Brigade; the novels were a cult phenomenon among Romanian teenagers. In the second story in the series, the kids embark on a summer adventure that leads them to a mysterious old castle reportedly haunted by a girl dressed in white. Driven by their innate curiosity and the allure of the unknown, the friends explore the castle, finding clues that hint to a hidden treasure within the grounds. These stories still hold a special place in my heart, yet when I think about them today, I wish the kids could have used AI to expand their knowledge—not by accessing the sort of digital libraries we have focused on thus far, but by using AI to extract knowledge from built and natural environments.

If my heroes had enjoyed access to a pair of smart glasses embedded with cameras, a processing unit, and an AI tool that allowed them to scan and analyze the artifacts and paintings in the castle, they'd have acquired instant historical context. The AI glasses would have strengthened their code-cracking abilities and situational awareness and helped them beat their rivals to the treasure. In one of their outdoor escapades, the brigade encounters potentially dangerous wildlife. The glasses could have helped them identify the species, its habits, and what threat it posed. The kids often camped outdoors, too, and while setting up for the night they might have benefited from an AI-equipped drone that could scour the terrain, suggest the most secure and comfortable spot, and perhaps even help in identifying edible plants or potential water sources nearby.

Although I am not quite as adventurous as these heroes of my childhood, and have dedicated much of my career to work in the laboratory, I seek out ways to deploy tools in nature to enhance our knowledge of the wilder world. For one project we used AI to broaden our knowledge of cows. There is comparatively little behavioral research on these incredible animals, relative to wild species, because they are so domesticated. Many cultures and companies treat them more like miniature biological manufacturing plants than intelligent creatures. A few years ago, my students and I worked with a scientist at the U.S. Department of Agriculture, Dean Anderson. We designed small GPS tracking devices that a population of domesticated cows could wear on their collars, then fed this data into a machine learning model to track, optimize, and predict their movements and activities. The findings were fascinating. We learned that cows sleepwalk, for instance, and that they may exhibit what we think of as babysitting, with members of the group caring for calves while the parents focus on other duties. The model also revealed that cows tend to cluster in groups with specific leaders. The work provided Dean and his colleagues with new knowledge about when the cows wanted to eat, roam, and rest, and generally created a richer picture of these creatures and their surprisingly complex social lives.

There is increasing evidence that AI will bolster our knowledge acquisition beyond our planet, too. Today, astronomers rely largely on robotic telescopes based on the ground or orbiting in space, where their views are clear of natural and artificial atmospheric interference. Think of these images not as pictures but as data and information, and AI as a way to observe, study, and find hidden patterns, to function as a telescope for knowledge. One of my favorite examples is the work of the Yale University astrophysicist Priyamvada Natarajan (known to her friends and colleagues as Priya). An expert on black holes and the mapping of dark matter, the mysterious and invisible material that appears to make up most of the mass in the universe, Priya has found a very interesting way to put AI to work in her field. Astrophysicists studying black holes and dark matter rely heavily on simulations, which serve as proxies for laboratory experiments, as they cannot perform the sort of controlled experiments common to other scientific fields. So, to test theories with observed data, they use simulations of a patch of the universe. We can detect distant quasars and black holes—the cosmic maws that swallow up all the light and matter in their vicinity—but because we are looking so far into the distance, we are really only "seeing" the brightest and biggest objects. This is fundamentally limiting, as these are the rare and extreme outliers. Basing our understanding solely on the brightest objects is the equivalent of studying American history only by researching the presidents.

To attempt to work around this problem, astrophysicists use data from a patch of our very large universe that we know very, very well and build simulations based on theoretical models to reproduce the observations. However, to match the larger set of observed but inherently incomplete data from the cosmos, these simulations need to be scaled up in volume. In doing so, astronomers get a sense of the smaller, dimmer, or even invisible objects that *should* be out there, but are too faint to be seen by direct observation. What Priya and her colleagues did was train a machine learning model and apply it to populate the larger, scaled-up simulations. She used AI to uncover hidden patterns in existing data, and the work revealed that we seem to be missing an entire population of black holes.

AI has also played a pivotal role in the groundbreaking discovery and analysis of gravitational waves, the ripples in spacetime caused by massive celestial events like the merging of two black holes. To detect the waves, scientists need to analyze vast amounts of data collected by instruments such as the Laser Interferometer Gravitational-Wave Observatory (LIGO). Picking out the telltale signal of a gravity wave is like trying to hear a faint whisper on the far side of an arena during a Taylor Swift concert. In this context, AI functions as an expert sound engineer able to tune out the noise and amplify the whisper. The technology can quickly pick out the unique patterns that indicate a gravitational wave, and even estimate details about the event that caused it, such as the size and spin of the merging black holes. Additionally, AI can help transform this information into knowledge by simulating what these waves should look like based on different theories, helping scientists confirm their models.

• • •

What all these astronomers, gardeners, lawyers, students, and adventurous fictional teens have in common is a powerful desire to acquire quality knowledge. This will very naturally extend to business and industry as well. By layering sophisticated AI models atop the collected wisdom of human experts, we can foster a collaboration between human and machine that will have practical applications which extend far and wide. Many companies are now developing technologies for summarizing manuals and other large information repositories. For example, Oracle uses AI for document summarization in its Enterprise Resource Planning (ERP) Cloud platform, helping users quickly understand the content of financial reports, invoices, or other documents. In manufacturing, predictive analytics powered by AI can analyze machine performance data to forecast maintenance needs, reducing downtime and improving overall equipment effectiveness. Similar AI tools will help logistics specialists forecast shipping delays and potential disruptions in the supply chain by effectively swallowing more knowledge, considering factors such as weather patterns, customs regulations, geopolitical events, and historical shipping data. This allows businesses to plan ahead and make informed decisions, mitigating risk and ensuring that goods arrive on time. Retailers will be able to learn and ultimately know more about their actual and potential customers, as AI analytics will help them understand purchasing patterns and preferences, allowing for more personalized customer experiences and more effective product placement. There are numerous applications in healthcare as well. AlphaFold alone has vastly expanded our knowledge of the proteins critical to human disease. And we will certainly benefit as patients, too. AI can analyze individual patient data to create a comprehensive medical history, suggest tailored treatment plans for healthcare providers to consider, and enhance providers' understanding of individual responses to therapies. Our farmers could even grow higher-quality food, as

AI-driven analysis can provide them with detailed information about weather predictions, soil health, and crop condition, allowing for more informed decisions about planting and cultivation. For example, the startup company ClimateAI is already helping farmers select the most resilient crop types and seeds given the local climate predictions.

These applications suggest that we might eventually develop subject- and industry-focused AI librarians—virtual digital experts that can directly generate quality knowledge hidden within the vast array of information in the digital, built, and natural worlds, or uncover patterns that point to new knowledge, which would then be curated and fine-tuned to each task. Scientists could use virtual research assistants to pick out relevant papers among the hundreds of millions of published works, then have those highlights synthesized into easily readable summaries. Business leaders could use virtual research assistants to extract on-demand task-related knowledge, enabling informed data-driven decision-making processes and more precise and informed strategies tailored to their work. Salespeople could rapidly become experts in a new industry before critical meetings with prospective customers.

Generally, AI can transform vast and chaotic seas of information and map out a tailored knowledge journey for each of us, no matter your interests or occupation. The technology could help us transition from a state of information abundance to one of knowledge evolution, where learning is continual, adaptive, and deeply personalized. So then the question becomes how one might carve out a niche or a professional pathway in this changing world. I suspect those who learn how to acquire knowledge quickly with the help of AI, and parse the results intelligently, will have an advantage over those who rely on old-fashioned information retrieval. Or perhaps the individuals who depend on classic studying and knowledge acquisition techniques will distinguish themselves because they store more information in their minds and don't need to turn to a tool for assistance.

This reminds me of a story about a friend of mine. In Communist Romania, during the 1980s, we were not allowed to travel outside the country's borders, so we dreamed of visiting the grand cities of Europe. One of my friends became so entranced by the idea of Paris that he memorized a map of the city from a book. Years later, when he finally had the chance to visit, he knew every neighborhood and street, and could navigate Paris like a local. He enjoyed the visit more because of the knowledge he'd amassed in his mind. This is a reminder that we should not drain our brains of facts and details and rely solely on intelligent tools. There will still be tremendous value in memorizing maps, reading entire books or shelves of them, and generally filling our wonderful minds with knowledge. What we know can impact what we do in the future in surprising ways. Steve Jobs credited the elegant fonts he introduced in various Apple products to the calligraphy courses he attended in college. Yet we cannot memorize every map or read every book, and this is where AI will play an increasingly important role as a knowledge acquisition tool. So I would suggest becoming adept with both approaches. And then what? Well, unless you aspire to win a trivia game show, accessing and acquiring knowledge is not enough. It's what we do with this information, and what ideas we generate from it, that really matters. What we really want is to turn this knowledge into insight.

3

Insight

A FEW YEARS AGO I WAS DELIVERING A LECTURE ON SELF-driving cars to a company in India when my host asked me a surprisingly difficult question. This executive had been instrumental in arranging and facilitating my trip, and he was the first to raise his hand when I finished my talk. Naturally, I assumed he was going to query me about something related to AI or autonomous robots. Instead, he stood and asked, "How do you explain insight to a seven-year-old child?"

The question stunned me. I stumbled through my reply, and I couldn't stop thinking about it during the rest of the Q & A, my time in India, and my flight home. What is insight? The standard definitions call it a deeper understanding of a concept, person, or thing. In this sense, it is a level up from knowledge in that hierarchical pyramid. Or perhaps it is the engine that advances us from knowledge to wisdom. But how to explain it to a child? Well, what do most children like? Dessert. Ice cream in particular. Yet they probably don't think about where it comes from. So, if I were explaining insight to a young mind, I might teach them to connect the vanilla scoop atop a cone with the cow in the pasture, the original source of the cream, or even the grass in the field that provides the animal with the energy and fuel needed to produce its milk. These are completely disparate entities. The cow, the grass, and the scoop of vanilla ice cream appear to have very little in common. Yet they are directly linked. Insight, in this context, is about uncovering connections that one didn't realize existed.

To be clear, I do not think we should outsource the uncovering of insights to AI. This is one of the true joys of learning, especially learning broadly, and it is the wellspring of many human achievements. But AI could be incredibly valuable as an assistive insight engine, for a few basic reasons. First, our current AI and machine learning tools are designed to ingest and find patterns within massive datasets. Second, AI models "think" differently than humans, so it is likely they will uncover patterns and connections that we might not see or notice. Third, these tools now have the ability to generate or synthesize new ideas, making suggestions that we can act upon; they can uncover and highlight insights that might be worthy of consideration. The potential applications do not need to be limited to business strategies or grand challenges in astronomy or biology, either. We could all benefit from uncovering hidden insights relevant to our daily lives, our work, even our leisure and play.

But let's look at science first.

In 1962, the historian of science Thomas Kuhn published *The Structure of Scientific Revolutions*, in which he details a novel theory about the evolution of scientific knowledge. He explained that while science often progresses through the steady accumulation of knowledge, where ideas build upon one another in a gradual manner, this linear progression is punctuated by massive, disruptive changes called paradigm shifts. These are moments when an entirely new idea overturns the established view and reshapes the entire scientific landscape. The geocentric view of the universe was overturned by the heliocentric conception that replaced the earth with the sun as the center of our local cosmic neighborhood. Newton's conception of gravity held for hundreds of years until Einstein overturned it with his general theory of relativity. Suddenly we moved from the Newtonian view of the world to a very different Einsteinian conception, in which gravity changes the very fabric of spacetime.

The shifts in thinking that Kuhn detailed in his book have parallels in nature, too. Originally scientists believed that evolution was a steady process. Organisms evolved through the regular accumulation of minor adaptations. But evolutionary biologists have since learned that species tend to experience long periods of stability punctuated by rapid changes in relatively short timespans. Everything is proceeding along steadily until there's a massive shift. A new creature is born with a radically different gene, perhaps, that renders it more fit for the environment, and this gene spreads. The combination of linear progression with sudden, transformative shifts offers a more nuanced and complex understanding of both scientific and biological evolution.

There will be no virtual version of Einstein or Newton, but AI will be very, very good at supporting the gradual progression, or the incremental steps between the paradigm shifts. We are already starting to see this today. The MIT physicist Max Tegmark and one of his graduate students, Tailin Wu, developed what they dubbed an AI Physicist—an intelligent agent that proved capable of studying simulated universes and extracting the theories or laws that govern those imaginary worlds. The AI Physicist is like a digital detective that creates and tests theories about how the world works. Imagine a team of mini-scientists, each with their own idea about the rules of a game. As they watch the game unfold, these mini-scientists adjust their ideas to better fit what they observe. The AI Physicist acts as their leader, deciding which mini-scientists have the most promising ideas and letting them work together to develop a more complete understanding of the game's rules. This approach mirrors how human scientists build and refine theories about the world, but the AI can do it much faster. Tegmark's tool has been used to discover new rules in simulated environments and holds the potential to help us better understand the natural world.

My friend Hod Lipson,* a Columbia University roboticist, is doing similarly fascinating work, exploring how we might use AI to generate new scientific insights. Hod and his students are looking to develop AI engines that can independently generate scientific models or hypotheses when given a large enough body of data. In one case, a physicist at the Conseil Européen pour la Recherche Nucléaire, or CERN, provided him a dataset in the hope that Hod could solve a challenge relating to atomic nuclei. Hod and his colleagues ran the data through their AI hypothesis engine and generated an equation designed to express the patterns inherent in that dataset. Excited, he sent the result back to his contact at CERN. Yet the physicist was disappointed. He informed Hod that the equation he sent was already known.

Meanwhile, Hod was thrilled! The equation might not have

* I realize that I refer to multiple researchers as my friends, and that this might become tiresome to some readers, but I can't resist, as I count scientists like Hod among my closest colleagues.

helped the physicist, but it validated Hod's hypothesis engine, as it had generated a proven equation entirely on its own. Later, another scientist contacted him about a problem in cell signaling, and once again the model extracted patterns in the dataset provided and generated an equation to describe them. In this case, though, the scientists couldn't quite explain why the equation worked or how it arose from the data. Hod compared it to copying someone else's homework. You might get the right answers, but if you don't know how you got those answers or how to explain them, you have not truly learned anything. He has emphasized that as AI moves forward, we will need to understand why these models produce what they produce—how they got the correct answers on the test.

While this example led to consternation instead of clear insight, it reveals the incredible potential of AI in science. The universe is still shrouded in mystery. Most of the mass in the cosmos is invisible. Our galaxies are drifting away from one another powered by an equally mysterious dark energy we cannot explain. We don't merely need new and more advanced telescopes and microscopes. We need new tools to help us make sense of the vast amounts of data they gather. When Hod released his work, a prominent news organization suggested this might be the first step toward the obsolescence of theoretical physicists. This missed the point entirely. Hod used AI as a tool. This tool produced an insight in the form of an equation to describe a previously inscrutable dataset. But without scientists to make sense of that equation, the underlying variables or constants, and how to apply them to our larger understanding of the natural world, the AI's product is relatively meaningless.

We need people more than ever.

* * *

As a scientist I'm prone to associate insight with the natural world or complex engineered systems, but the business world depends heavily on insights. Raw data points are transformed when we assign values and meaning to them. Once we know that they are monthly sales figures, they become information, and after a business analyst studies them to uncover patterns regarding peak selling periods, we have graduated to knowledge. That analyst advances to insight when she finds that certain sales spikes correlate with an additional set of information and knowledge, such as a particular marketing campaign or socioeconomic trend.

Many businesses have already begun using AI in this way. The company Brandwatch, for example, has long offered a technology solution that tracks mentions of a company or a salient topic on social media. Traditionally, the solution's users had to sift through this captured data one message at a time. Then Brandwatch developed a way to use AI to generate summaries and extract the key themes. Instead of reading through 1,200 social media posts, for example, a user can ask their GPT-powered Brandwatch tool to generate four or five key themes distributed across those posts, then focus their mental efforts on developing a way to respond or react to these trends. Another machine-learning-enhanced Brandwatch solution looks at the collected mentions of a particular company, pulls out popular topics, and then segments these topics by demographic. In one case study, while an older generation tended to show more interest in a popular automotive manufacturer's new factory, the Gen Z crowd talked more about a car's performance. These tools accelerate the speed of work for business analysts, as they save them the time of browsing the collected data to extract the same information, but they also enhance the analyst's ability to extract insights as fewer hours are spent weeding through data. The AI uncovers the connections within different customer groups, and the human analyst can spend more time considering those findings and developing strategic responses.

• • •

What about insight into ourselves? Sleep is a prime target for exploration. There are any number of apps that will monitor and compile basic data gathered from your phone or smartwatch to help evaluate how you slept the previous night, and scientists are beginning to use AI to untangle some of the mysteries around more serious sleep disorders. They can do this now because they have enough quality data—a prerequisite for AI and machine learning. Sleep experts have long used polysomnography tests to collect data on a patient's blood oxygen levels, brain activity, breathing, eye movement, heart rate, and more during sleep. Traditionally, a patient spends a night in the hospital or sleep center, and experts then interpret the resulting data independently to understand a patient's particular case. They might determine that the patient has a particular type of apnea, for example. The Stanford University sleep scientist Emmanuel Mignot and colleagues have shown that AI models can review this data and score or diagnose a patient as effectively as a human expert. When he thought more about the possibilities, though, Professor Mignot realized this was only the beginning, and that AI had the potential to tell us something new that would be helpful in ways we didn't expect.

Soon Professor Mignot began using AI to explore whether there might be some connections between sleep patterns and various diseases. One of the first insights his model uncovered was a link between certain sleep patterns and narcolepsy. In another example, the model identified very specific behaviors during different sleep stages that were correlated with Parkinson's disease. And Professor Mignot and his colleagues suspect there may be many more such connections to discover, including early signs of heart attacks, Alzheimer's, or even changes in mortality risk. He envisions this technology becoming more broadly available, so that instead of visiting a sleep center and being subjected to a rigorous polysomnography test, the average person could use the gadgets that are increasingly becoming part of our daily lives to measure biological vitals. Once a month you could have your movement, breathing, and heart rate monitored as you sleep, and an AI model could scour the data for patterns that warn you if something looks troubling or your risk of developing a certain disease or condition has increased.

My MIT colleague and friend Professor Dina Katabi has done amazing work in this area as well, and believes that sleep can serve as a mirror for various health conditions. Indicators such as early rapid eye movement during sleep stages may hint at depression, and interrupted slow-wave sleep might signal the onset of Alzheimer's disease. She has developed a non-invasive, almost magical way to monitor a person's breathing as they sleep in order to reveal early-stage Parkinson's disease, now the world's most rapidly proliferating neurological condition. Current methods often diagnose the disease when 50–80% of the damage to the brain has already occurred; with AI, the diagnosis is sooner. Dina's novel technical solution monitors ambient radio signals for disturbances that correlate to movements or changes in breathing. Using machine learning, her wireless system can discern sleep patterns that correlate with signs of specific diseases or conditions. The technology has demonstrated remarkable potential, achieving up to 90% accuracy in preliminary findings. Imagine if systems like this were active all the time. Early and accurate diagnosis would become the norm rather than the exception. Our checkups would not be annual, but almost constant, and AI systems would alert healthcare professionals when an early warning sign or alarming trend appears.

Dina's diagnostic results are built on a system called Emerald, which is a sort of high-tech, invisible stethoscope that uses WiFi signals to monitor your health. These signals move all around the room, and as you move and breathe, or as your heart beats, you cause subtle changes in the way the signals bounce back. Emerald sends out WiFi signals, captures their reflections off your body, and analyzes the tiny changes in the signals to track your vital signs and movements. The device is effectively listening to the echoes to figure out what's happening inside your body, allowing for continuous, non-invasive health monitoring. Emerald also exemplifies the power of AI to analyze complex data that comes in a multitude of forms—in this case, different signals—by providing real-time insights into a person's health and well-being that would be challenging to discern through human observation alone. This approach is particularly valuable for keeping an eye on elderly people or those with chronic conditions, as it can provide real-time insights into their health and alert caregivers in case of an emergency, such as a fall.

• • •

There are a few ways I'd like to improve our insight-generating AI engines. Today we typically need to pull information, knowledge, and insights from AI, especially with language models.

You need a good prompt, which in turn requires a knowledgeable person to "hunt" for the insight. If the prompt is bad, the model is not going to give you the best answer. What I'd like is the technological equivalent of a really smart person who is going to spend their life studying some particular domain and then coming up with insights. Think of the Mentats from Frank Herbert's famous *Dune* novels. These characters were pattern recognition and insight generation machines who served the powerful emperors and dukes. Their advice enabled their superiors to craft economic, sociopolitical, and military plans and strategies, and revealed the veiled intentions of their enemies. Similarly, we could develop virtual Mentats that push insights out to us as opposed to waiting for us to extract them through carefully crafted prompts. These models could actively suggest to people working in a particular area that they think about X or Y, because the tool learned from all these patterns and all this data that there was some exciting potential there. Then the user could investigate the insight directly. In the world of *Dune*, the service of a Mentat was reserved for the rich and powerful, but this kind of tool could be made easily available to all of us.

The challenge of extracting valuable insights from AI, and shifting from a pull-based interaction to a more intuitive push-based system, is a fascinating frontier that has concrete applications across diverse industries and individual tasks. Analyzing production data in manufacturing facilities can reveal hidden inefficiencies or bottlenecks, leading to strategic changes that optimize overall performance. Products could be delivered to their destinations faster, as AI will be able to scour historical shipping data to uncover patterns common in delays, leading to improved practices and strategies. In healthcare, AI-driven analysis of patient data can uncover correlations between symptoms, treatments, and outcomes, potentially leading to innovative treatment strategies. Analysis of energy consumption data can provide insights into usage patterns, informing conservation strategies and policy decisions. Applying AI to crop data could help uncover correlations between various farming practices and outcomes, guiding future farming strategies. Name an industry and there is very likely a valuable application for AI-driven insights.

As I write these words, the field is advancing at a phenomenal pace, so it is reasonable to assume that we will have ever more capable tools or even virtual Mentats very soon indeed. When these tools do become available, and we can interact with our personal automated insight generators, the question then becomes what we might do with these findings. Of course, that depends on your field or industry. Personally, I'd hope these insights would be inspiring enough to kickstart a new project or company. To me, insight is merely the beginning. The next stage? Creativity.

4

Creativity

IN NOVEMBER OF 2022, THE TURKISH-AMERICAN NEW media artist Refik Anadol launched *Unsupervised*, an exhibit at the Museum of Modern Art (MOMA) in New York City. The titular reference to one of the common forms of machine learning* was no mere attempt to capitalize on the AI hype. Refik and I met at a conference, and I was fascinated by his unique creative process. He confessed that he has been intrigued by the idea of intelligent machines acting as our friends since he was a child, when a neighbor brought home a VHS cassette of the science fiction classic *Blade Runner*. One of Refik's cousins spoke English, and he was tasked with translating the dialogue in real time as the family watched the film. The movie enthralled Refik. Never mind the dystopian setting—he was entranced by the vision of intelligent machines.

* Unsupervised learning is a common approach to advancing machine learning systems in which the solution can improve its performance over time without human intervention or input.

Later, he began teaching himself how to program, and by 2012, he conceived of a very interesting idea. Digital artists had long been creating works from pixels, but Refik wondered what might result if these pixels had some intelligence of their own. What if the pixels responded to one another and even to outside sensory input? He coined the term "data painting" for this new approach, and his projects became increasingly sophisticated in the years that followed, culminating in *Unsupervised*. To create this work, Refik experimented with multiple generative AI tools before selecting one that best suited his aesthetic. With the help of MOMA curators, he spent six months training his chosen AI model on 380,000 images representing nearly 200,000 classic works in the museum's collection. Once the AI had been trained on the art, Refik and his team installed a two-story-tall, high-resolution video display on the museum's ground floor. Next he linked the display to a camera in the ceiling, which tracks the movement of people below, and connected it to a local weather data feed. Changes in the weather and the motions of museum patrons alter what happens on the screen. The pixels respond to the stimuli.

Refik compares the resulting exhibition to an AI dream of MOMA's collection. Of course, the system doesn't actually dream like a human; the images were created by a particular type of AI algorithm known as a Generative Adversarial Network, or GAN. This algorithm is unique in that it pits two different machine learning models against each other. The computer scientists who pioneered GANs in 2014 likened the competition between the models to a game between counterfeiters and the police. The former works to dupe the latter, but competition forces them both to evolve and improve. So as one model generates an image based on the training data, the other tries to guess if it's actually from the data or a clever impersonation. As they iterate, the image quality improves. The particular GAN model that Refik and his team selected allowed for more control over the images, enabling him to further shape the work.

The exhibit was an instant sensation. Crowds responded adoringly, moving with the images, and despite some critical backlash MOMA extended the installation repeatedly before making the work part of its permanent collection. While some of the criticism was related to the use of a controversial technology and resulting questions about who truly authored the work given the central role of an intelligent model, I find Refik's story to be incredibly inspiring, and not only because I am drawn to a classic American immigrant success tale. Refik found a way to use AI to move people through art in a new way. In doing so, he produced something that is not only beautiful but wonderfully strange, and demonstrated the potential of AI as a tool to both support and enhance human creativity. I've long been fascinated by the interaction of people and art. In 2018, for CSAIL's entrance in the Ray and Maria Stata building at MIT we commissioned an installation by another pioneer of AI art, Karl Sims. The piece uses feedback from a Kinect system to produce beautiful, dynamically changing images that capture and represent the movements of the passersby.

• • •

What is creativity? In its most fundamental form, I think of creativity as the ability to conceive something new and original, whether it is an idea, a solution, an artistic expression, a business plan, or a scientific hypothesis. This spark of innovation is a defining trait of human intelligence, an innate capacity to step beyond the bounds of known and habitual patterns and venture into the realm of the unexpected and uncharted. It's a process that involves a synthesis of existing knowledge, keen observation, imaginative thinking, and, sometimes, a healthy dose of serendipity. This dynamic amalgamation of elements leads to the emergence of novel ideas and works.

There is an interesting parallel here with the concept of punctuated evolution in biology—the brief, intense periods of change that lead to adaptations capable of redefining a species. Similarly, progress in human societies does not typically follow a steady, linear trajectory. Rather, it is marked by moments of intense creative output—paradigm-shifting innovations, breakthrough ideas, revolutionary art and literature—that redefine the course of history. Just as the flux of information and ideas can trigger creative punctuations in human societies, AI can act as a creative catalyst, facilitator, and amplifier. With Refik Anadol's breathtaking visual installations, AI's capacity to process and interpret mammoth amounts of data has provided a new type of pigment—the data—and a new type of brush—the algorithm. At the same time, though, AI can be an additional brush in the artist's studio, a tool that can help creators in fields as varied as filmmaking and scientific research play with ideas and concepts, make their visions real, add unexpected qualities or dimensions to enhance their work, or even simply help them through the rough patches in their processes by functioning as productivity aids. This use of AI as a creative tool does not need to be limited to world-renowned artists. Anyone can use these tools to enhance their creativity. Even kids.

As a child, I'd often get into trouble for drawing characters and scenes on my bedroom wall. My parents would make me promise that I'd never do it again. Invariably I'd find myself inspired, forget or disregard my promise, and grab another marker and set to work again. I understand my parents' perspective; they wanted clean walls! Yet if I could travel back in time, I'd love to transform that plain surface into an AI-enhanced smart screen, one that could scan my simple sketches, build animated models of the characters, and bring them to life on the wall. By adding a microphone, speakers, and some sensors, these AI-infused characters could even interact with my childhood self, engaging in simple conversations or even playing games.

Here AI would have enhanced my initial drawings, adding a dimension I could not have created on my own, but in other youthful artistic endeavors, I could have benefited from AI as more of a creative assistant. My cousin Anca and I loved to write scripts and stage puppet shows. Our longest play was called *The Little Girl Who Could Not Finish Anything*. The title was a reflection of our frustrations, as we never quite completed what we set out to write. Here we might have used a large language model to help us through writer's block or perhaps suggest possible endings to our plays. The little girls who could not finish anything would have benefited greatly from using AI as a tool in their playwriting process. To be fair, I don't think this would have changed the arc of our lives. Anca and I wouldn't be dramatists today instead of our current roles as a physician and an AI researcher, respectively. Yet it would have been wonderful to have finished our work.

These aren't merely tools for future or current Mozarts. Anyone can use AI to be more creative. A young parent with an idea for a children's book could enlist different tools to generate a finished product for their kids within a few days or weeks, instead of spending years honing their work in hopes of attracting a publisher. The result might not be a masterwork, but it would be a meaningful book for that family.

• • •

Many artists and creative professionals are naturally intimidated by such AI tools. Some are rightfully offended, too, given that most generative AI models were trained on publicly available digital art, which means it is likely that these tools acquired their skills at least in part by ingesting and studying patterns in the works of others. Yet artists have been doing exactly that for centuries or more. The philosopher Joanna Zylinska, writing in the journal *Science*, noted that Salvador Dalí studied the art of the masters and effectively remixed them to create a style all his own. Similarly, the singer Tony Bennett once remarked that listening to one performer might lead to theft, but studying and borrowing from all of them was research.

Technology has always changed art. The Adobe designer Aaron Hertzmann has drawn enlightening parallels between what is happening now and what took place in the late nineteenth century with the introduction of photography. At the time, the painter J. M. W. Turner declared this shift to be the end of art. Yet art endured, and painters created Impressionist, Post-Impressionist, Cubist, and other profoundly creative works of art that delight us today. The skillful application of paint to canvas is still a beloved medium of expression. Photography evolved as its own unique form, and its facility for capturing real scenes shifted the focus of painters away from realism as the ultimate goal. The use of photographic technology did not kill art. In fact, it might very well have pushed artists in entirely new directions and encouraged them to abandon realism and experiment with abstraction. The camera may have led to Cubism.

One of the interesting points about Refik Anadol's work is that people didn't dwell much on the particular GAN used to generate the images. In addition to the larger conversations about the role of AI in general, they talked about Refik. He was the one who conceived and implemented the idea. He tested and tuned the models, curated the data, and tied the model's output to external sensors. His creative fingerprints are everywhere. As viewers and consumers of art, we value seeing the imprint of the artist in the stylistic tendencies of a writer's prose, the brushstrokes of a painter, or the particular design and implementation choices of someone like Refik. We want to see the human influence in the work because it helps us connect with the artist. These tools are not replacing artists. They are adding new dimensions to their creative output and enhancing their productivity.

• • •

We have been experimenting with AI as a creative tool in my lab, only instead of making art, we're designing robots. For many years roboticists merely imitated the human body or put mechanical boxes on wheels when conceiving new machines. We gave our robots humanoid arms and legs, and this limited us as a field, because it forced us to operate inside a very fixed and limited vocabulary of shapes and forms. It's a bit like telling an artist they can *only* work on canvas. So in my group we decided to break through these limitations by using diffusion, one of the algorithms that enable generative AI, to see if we could develop more interesting and capable robot bodies. Our plan was to design a system that could optimize the shape of a robot and its function at the same time.

The great artist Michelangelo supposedly created his masterwork *David* by starting with a block of marble and carefully removing everything that was not the biblical hero.* Similarly, our diffusion model for robot bodies takes a random block of particles, removes the excess, and fine-tunes the remains into a specific shape. This approach could easily turn out all sorts of unusual robots, but there is no guarantee these machines would function effectively in the real world if we were to build them. So we tweaked the diffusion model, forcing it to optimize its chosen shape for a specific function, such as crawling, balancing, or hurdling an obstacle. Then we ran it through trials of the assigned task in a simulation engine, a sort of virtual space governed by the basic physical laws of our world. (The DiffuseBot system also determines the robot's stiffness at various spots, and where the artificial muscles and sensors should be placed.) By testing the designs in simulation, the system could learn what worked and failed, and rapidly iterate through new forms until the resulting virtual robot accomplished the given task and satisfied our specifications. In the end, the winning designs were marvelously unusual. One looked like a gummy bear. Another had four legs connected by a worm-like structure. We turned one of these virtual robots into a gripper by following the system's instructions. Our odd creation worked and we built many new robots.

This was science, and not necessarily art, yet the work demonstrates the role of AI in fostering creativity. The diffusion model did not replace the human element. It augmented our work, and we should all think about using AI as a tool that can help us dig deeper, reach further, and imagine more boldly across all fields

* I say "supposedly" because the quote attributed to Michelangelo has not been verified.

and industries. With AI by our side, we might find ourselves on the brink of a new era of creative punctuations, driving progress in exciting and unexpected directions.

• • •

The generative AI models that create images, videos, poems, and songs are not artists. A great writer does not copy. A great writer is copied. Yet these tools in general will help people express themselves in new ways, explore new creative avenues, or simply work more efficiently. Existing artists, innovators, and inventors could use AI models as thought-provoking or creativity-enhancing aids, or as additional technologically enhanced brushes in the studio. Other little girls like Anca and me will actually finish their plays. In short, while I understand the fear, we shouldn't think of AI as competition. We should look at AI as a way to help us all be more creative in work and in life.

In early 2023, for example, a few of my students presented me with an extraordinary gift. Although I'm a lab director and active researcher, I'm also a mentor and teacher, and in the world of robotics and AI, our students don't thank their instructors with apples. Whimsical inventions or clever technology demonstrations are more appreciated, and my students Alexander Amini and Ava Soleimany presented me with a series of portraits of myself. Normally I might find this strange, but these particular works of art were created by an AI model they'd designed and built just for this task. The students developed a tool similar to Stable Diffusion, popular AI programs that generate artworks in various styles. The results are stunning, the portraits striking, but what really impressed me was not the model but my students. These young engineers had transformed themselves into artists through the clever application of code. They developed a program capable of expressing their gratitude, much as a painter would use a canvas and brush.

A very different case involves the rapper Lupe Fiasco, who spent an academic year at MIT teaching a class on the history of his art form. He is an accomplished, established writer and performer. He doesn't need the same sort of help as a pair of computer scientists trying to create an artistic portrait. Yet he still wondered if AI might be able to offer him some assistance, so he collaborated with Google to develop a few tools. Initially, his partners at Google thought he would want something that could write his raps for him, but Lupe resisted. He didn't want a technology that would remove the artist from the creative process. He wanted something to empower the rapper. So he collaborated with a team of Google engineers to develop a collection of ten different AI tools that help him optimize his creative workflow. One takes a word as input and spins it out into similar-sounding words and phrases. The model transforms the prompt "coffee" into "cup of tea" and "coffin fee" and other variations, for example. Lupe didn't look at these tools as competition. To him they functioned more like a new instrument; he envisioned confining the role of AI to specific stages. Lupe wanted to remain in sole control of the initial ideation, yet he welcomed AI as a tool to help during the production and refinement phases of his writing.

The visual artist Steve McDonald offers another intriguing example of how creators can weave AI into their workflow. McDonald is best known for his adult coloring books, which feature intricate line drawings of real and imaginary cities and have sold millions of copies worldwide. He's a classically trained artist who spent many years, as he put it, pushing oil around on canvases to see the results up on gallery walls. Yet McDonald has been embracing technological tools since he started professionally more than thirty years ago. He has experimented with projectors, airbrushing, digital media, and when he first heard news of generative AI solutions, he began testing them immediately, and eventually developed a creative process that has effectively added multiple finely tuned AI engines to his palette.

The work begins with an original idea or question. At one point, McDonald was thinking about old maps, and how they used to have regions labeled "here be dragons," and he imagined worlds in which dragons were normal. He started by sketching a few dragons with his digital stylus, then fed a dozen of these hand-drawn images to a custom diffusion model and created his own dragon-generating AI. He selected a hundred of his archived drawings of villages that he liked, and trained a separate AI model on those. Next he had the two models generate a series of images of dragons lurking in the background of one of his imaginary villages. These weren't exactly *his* sketches. The AI models generated variations based on his style. He selected one of these hybrid images, moved it into Photoshop, then adjusted and touched it up in various spots. After these edits, he ran the work-in-progress through another AI model he'd developed that tweaked the color, and yet another that adjusted the texture of the image. All the while he was making edits and changes in Photoshop.

McDonald's process might involve a hundred steps and dozens of interactions with different AI models before he's finished. The result is a signature style unique to him, because it is based on a complex combination of his drawings and AI models he has trained on his work. It's his work, painted with a more complex collection of brushes. On the other hand, anyone who interacts casually with one of the popular diffusion models, and simply enters a prompt to generate an image, will likely end up producing work that is indistinguishable from the masses. As a large team of notable experts and computer scientists wrote in a 2023 paper on the subject, the programs have begun to develop their own aesthetic, with an apparent preference for hyperrealism, and creators have complained that it has become difficult to coax the engines into producing images in new or different styles. McDonald circumvented this issue by deploying custom AI models that he trained on his own art and particular preferences. "You can do more with these models than any tool we've ever had," McDonald insists. "The only thing holding you back is your imagination."

• • •

Although the potential is thrilling, I understand the fear. The changes will be disruptive. Some creative jobs may decline severely in numbers. At the same time, however, new ones will arise. This happens generally across industries—as my MIT colleague and economist David Autor has pointed out, 60% of today's jobs did not exist eighty years ago—and there is no reason to believe creative work will be different. The fear of machines displacing humans in the filmmaking process is a telling example, and a complex issue that stems from both valid concerns and misconceptions about the power and capabilities of AI. Recent advances in computer vision, natural language processing, and generative AI have resulted in systems capable of performing tasks such as scriptwriting, video editing, and special effects creation. However, these contributions from machines are best characterized as producing routine outcomes. The systems aren't writing inventive or canonical scripts. They are masters of cliché. While AI can assist with certain tasks within the craft of scriptwriting and enhance efficiencies in video editing, it cannot replicate the human creativity, cultural understanding, and emotional nuance that are essential to compelling storytelling. A study on creative product design work by the Boston Consulting Group backed this idea. The experiment split participants into two groups. Half had access to GPT-4 and the other half did not. The study found that while GPT-4 enhanced the creativity of individual participants, at the group level the tool produced 41% less variety in the output. The reason is that GPT-4 tended to suggest variations on the same ideas. The output lacked creative diversity and breadth.

While we must be smart about how we use these tools and how much we rely on their suggestions, there is cause for excitement here. If we return to the movie industry, a more productive and perhaps even harmonious partnership between technology and the filmmaking world could emerge through a collaborative approach in which humans and AI work together to achieve creative and technical excellence. An affordable educational component would need to be incorporated to help production teams acquire the skills to use these tools, but in the near future, AI systems could be creative assistants for scriptwriters, directors, artists, and others at various levels of the filmmaking process. These tools could help generate ideas, explore alternative plotlines and endings, or simulate visual effects. AI can assist with time-consuming and repetitive tasks in the filmmaking process, such as the simpler phases of video editing, color correction, and audio synchronization, and enhance filmmakers' ability to create storyboards by visualizing scenes, characters, and effects before shooting begins. This would help directors and production teams plan better, reduce shooting time, and achieve their artistic vision more efficiently. The director Ridley Scott is known for his intricate storyboards; moviemakers without a budget for concept artists could use AI-based storyboard generators. The future of filmmaking may be a synergistic human–machine process where AI takes on routine tasks, provides options, and makes it easier for the true artists to produce great work.

Yet that doesn't mean things might not get a little strange. Further into the future, AI could be used to create personalized experiences for viewers. During one recent extended holiday, my family binge-watched the second season of the popular drama *The Bear*. Although we were thoroughly hooked, we disagreed on whether we liked how the season ended. I would have enjoyed a chance to view AI-generated alternate endings, even if the video quality did not match the original. This could even open a new genre of film—interactive movie experiences in which viewers can project themselves into the film and interact with the objects and characters, influencing the storyline. This is not possible yet, but it is entirely conceivable.

I've focused mainly on the creative arts here, but the remarkable power of AI to foster creativity extends well beyond the realm of portraits, raps, and movies. These tools can be used to design new products with desired properties and functions, similar to our DiffuseBot work. The automotive industry could produce customized vehicles through AI-driven design and fabrication. We could utilize AI to design novel renewable energy systems that optimize the use of the power grid. In agriculture, I imagine AI-guided development of new foods that are faster to cultivate in vertical farms.

While AI could add dynamic layers to industries like manufacturing, fashion, and food production, we must acknowledge that there are aspects of creativity where the brushstrokes of artificial intelligence will never match the human mind. No artificial system will replicate the singular brilliance of human creativity. This assertion may sound commonplace, and perhaps even a little trite, but I do believe sincerely that there is something unique about our creative capacity, and I hope that more artists are inspired to experiment with these new tools as a means of furthering these innate capabilities and developing truly novel works, and that the rest of us use them to unlock the artist within.

5

Foresight

ONE OF MY FAVORITE MAGICAL ITEMS IN THE HARRY Potter series is the Marauder's Map, which allows the young wizard to track teachers, rivals, and potential villains. The map shows the grounds, halls, and hidden tunnels of his school, Hogwarts, and the footsteps of different persons of interest appear as they walk or sneak about. This proves to be a very useful tool for Harry during his various investigations and adventures. But it could certainly be improved. What if this map predicted where the characters were likely to go next?

In our research we have designed something similar for self-driving cars. Machine learning models are never perfectly sure of their predictions. There is always an element of uncertainty. What we wanted to do was reduce that uncertainty through learning and experience, so we developed a method called deep evidential learning. Imagine a random car driving through a city. At the start, the car's AI brain generates multiple predictions—say, ten—about where the vehicle will be at a given point in time in the future. This, in turn, produces ten different possible futures. Then, once the car hits that specified time point, the model compares the predictions to reality and makes adjustments to improve future predictions. Randomly chosen parameters, or best guesses, produce the initial predictions. So, to improve the predictions, it sharpens those parameters. In simpler terms, the model learns by acquiring evidence in real time and adjusting based on that evidence. The more evidence acquired, the more confident the model's predictions. Hence the name—evidential learning.

Ultimately, the car's AI system is not predicting the exact outcome so much as all the possible future paths and the likelihood of each one. Although we developed this model for autonomous driving, the approach is general and can be applied to any system in which knowledge and evidence are consistently acquired over time. For example, we have shown that it could be used in the chemical sciences to discover, design, and predict the properties of new molecules. We could adopt the approach to enhance Harry's magical map, too. In the world of the novels, the Marauder's Map had been tracking students and employees for centuries. If this map had been driven by an AI core, and if all this collected historical data were stored for analysis, then the AI variation of the enchanted parchment would have incredible predictive power. As characters moved, this AI-enhanced map would be able predict their likely path and ultimate destination based on the past history of others who'd traveled those routes, and readjust if they took unexpected turns. Harry would certainly have been able to elude the school's custodian, Argus Filch, with relative ease given the man's frequent and predictable nightly wanderings.

There is great potential for application in using AI to achieve foresight, which moves us a step beyond insight, as it looks at the road ahead, predicts likely future outcomes based on current data, and therefore helps us make more informed decisions about what to do next. Think of how the superhero Doctor Strange used his mystical time stone in the movie *Avengers: Infinity War*. In a critical scene, he rapidly envisions millions of possible outcomes in the superheroes' war with the villain Thanos. The difference with AI is that he would have been able to assign a probability to each of the millions of outcomes. I'm not sure this would have been helpful to the superheroes, though, as the low likelihood of success might have inspired them to give up.

But let's look beyond fictional and simulated worlds. Any complex system for which we have large, historically rich, and diverse datasets from a wide variety of sensors and sources could be a candidate for the predictive power of machine learning and AI. The mathematician Jim Simons built an investment firm on the use of machine learning to study data, identify patterns, and foresee fluctuations in the financial markets, and became one of the world's wealthiest individuals. I don't expect AI will be able to perfectly predict the stock market or the future, but it will give us valuable foresight into different possible futures, showing us what might happen with some degree of certainty so that we can prepare accordingly.

Consider traffic accidents, which kill more than a million people each year. Past efforts to predict crashes or identify high-risk intersections relied on poor resolution. The risk maps consisted of cells or pixels with a resolution of several hundred meters. Some maps used grids that were a kilometer on a side. A quiet street might be assigned the same accident risk as an adjacent stretch of an accident-prone freeway simply because they were part of the same oversized pixel. Ideally, a higher-resolution map would pinpoint a particular street corner or on-ramp as potentially fatal—not an entire neighborhood. My colleagues Mohammad

Alizadeh, Hari Balakrishnan, Sam Madden, and their collaborators, including researchers from the Qatar Center for Artificial Intelligence, reasoned that AI could be the secret to a new approach.

They developed a model that incorporates satellite imagery, road maps, data on past accidents, and information capturing the density and speed of traffic flow. The satellite images provided details on the number of traffic lanes and other fine details, including the density of pedestrians present. The historical accident data revealed where and when crashes had happened in the past. Ultimately, the researchers shrank the resolution of their map down to square pixels five meters on a side, trained an AI model, and found that it learned to associate specific road features and traffic flow patterns with higher accident rates. To gauge its effectiveness, they trained the model on US accident data in four major cities in 2017 and 2018. Then they tested it on accident data from 2019 and 2020, to see whether its predictions matched reality. To be clear, the model didn't pinpoint specific crashes. Predicting that a three-car accident would occur on a particular Wednesday evening in June, for example, would be impossible. But the model highlighted risky areas, such as on-ramps and intersections, that ended up incurring actual traffic accidents in the subsequent timeframe. This sort of foresight could be used by city planners to change the traffic rules in risky spots to reduce the chance of an accident.

If the model were trained on data from multiple cities, it could be generalized to other areas around the world, including those that might not have historical accident data. The model could inform those cities or locales of high-risk areas even without that information. Again, it won't tell you, as you approach an intersection, that there is a certain chance of an accident occurring right then and there, but it could encourage city planners to add or alter the use of stop signs or traffic signals at high-risk locations.

The value of foresight in healthcare is perhaps even stronger. We are constantly amassing more and more health-related data from electronic medical records, wearable devices such as smartwatches, genetic tests, novel approaches like the Emerald system developed by Dina Katabi, and many other sources. We can use AI and machine learning to analyze this data and extract valuable insights, pointing toward a more accurate, efficient, patient-centric and personalized healthcare system. My MIT CSAIL colleague Manolis Kellis and his collaborators leverage genetic data to predict the effects of certain variations in genes, enabling early interventions that may prevent the onset of diseases. Their work could help scientists understand the long-term impacts of genetic variations, pave the way for personalized medicine based on an individual's genetic makeup, and offer some foresight into the likelihood that a person might suffer from a particular hereditary disease.

There will of course be risks associated with greater reliance on AI models. My AI colleague Michael I. Jordan of the University of California, Berkeley, illustrated this point quite well in a 2018 Medium essay. Mike recalled how, a few years earlier, when his wife was pregnant, a geneticist informed the couple that medical imaging had revealed some troubling spots near the heart of the fetus—likely markers for Down syndrome. As it turned out, however, the study that suggested the link between those image features and the condition had used a machine with a very different resolution than the one this geneticist was working with. The troubling spots were actually just random noise.

In the field we are always working to reduce such false positives, but Mike's story is an important reminder that we cannot and should not rely entirely on machine intelligences despite their apparent clairvoyance. We should not let them make important medical decisions without steadfast human supervision. We have to work with these tools, incorporating our own deep knowledge and expertise. The work of my MIT colleague Dava Newman is a perfect example.

Dava, who is the director of the MIT Media Lab, has a very interesting professional history. She first gained widespread recognition for redesigning the traditional spacesuit into a more flexible, intelligent, but still protective garment more suited to exploration. She also circumnavigated the globe in a sailboat earlier in her career, and this wide range of professional and personal pursuits has given her a unique perspective. Dava has a deep understanding of our planet from both the cosmic and on-the-ground perspectives, and these vastly different viewpoints inform her new project to revolutionize weather prediction using advanced machine learning models. Dava and her team are enhancing the accuracy and reliability of weather forecasts through a novel AI system called the Earth Intelligence Engine. By analyzing vast troves of data from diverse sources—satellite imagery, atmospheric measurements, and historical weather data—this engine can identify intricate patterns and correlations that elude traditional forecasting models. The model can then generate more precise and timely predictions, enabling the anticipation of severe weather events, predicting their effects on infrastructure, and thereby allowing human planners to devise and implement ways to mitigate their impact. Plus, it works at both the global and the small scale, so it's capable of forecasting broad regional trends and high-resolution local events. In a different project, my research group used real-time data from a distributed network of sensors installed in rivers to provide local flood forecasting as an early warning system for remote villages and communities.

More accurate predictions of severe weather events such as hurricanes, tornadoes, and blizzards would allow us to save lives and property by providing authorities with more time to prepare and evacuate affected areas if necessary. Such capabilities would benefit various industries such as agriculture, energy, logistics, and transportation—all of which are deeply impacted by weather conditions. Farmers would be able to schedule planting and harvesting more efficiently. Energy providers would better predict fluctuations in demand, and shipping companies would be able to optimize their routes to account for weather. Airlines would be able to let you know that your plane is going to be rerouted or delayed long before you leave for the airport.

The Earth Intelligence Engine is a large project involving numerous contributors, but the link between Dava's experience and the ability of the model to work at both global and local scales should not be overlooked, as it offers another positive example of how we can leverage AI to further our own visions and works. Similarly, the oceanographer Raymond Schmitt of the Woods Hole Oceanographic Institute found a way to leverage AI and machine learning to verify his decades-old theory that clues to the likelihood of devastating droughts and floods might appear months in advance.

Generally, predicting seasonal events and trends such as these has been more art than science. Dr. Schmitt has long believed that the role of the oceans as a driving force in the global water cycle, or how water moves around and through the world, has been underestimated. The original inspiration for this theory was a devastating 1993 flood in the midwestern United States that pumped so much freshwater into the Gulf of Mexico that its salinity was significantly lowered. The astounding volume of water that caused so much damage in the Midwest, then drained into the Gulf, had to have come from somewhere, Dr. Schmitt reasoned, and he believed the ocean was the likely source.

Evaporation happens across the surface of our planet's oceans, and when seawater evaporates it doesn't carry its salt into the air. Freshwater rises up into the atmosphere and the salt is left behind. So, as more evaporation occurs over a given area of the ocean, the remaining seawater becomes saltier. Scientists have been measuring this variable—what they call sea surface salinity—since the late nineteenth century. Yet the data had not been used to forecast rainfall or weather patterns, and Dr. Schmitt believed that it might hold some predictive power. If the salinity of the water over a large enough patch of ocean increased significantly, that would suggest that an unusual amount of freshwater had been transferred from the sea to the sky. Depending on the prevailing wind patterns, he posited, this excess water might find its way to land and induce heavy rainfall or even floods.

In 2016, Dr. Schmitt and his colleagues published research suggesting that higher-than-average springtime salinity levels in a particular area of the Atlantic Ocean had led to increased rainfall during that summer's monsoon season in the Sahel region of Africa, which stretches across the continent south of the Sahara. Their idea wasn't to replace standard weather forecasting with salinity data, but to augment it and strengthen its predictive power. The group trained a machine learning algorithm on sea surface salinity data, along with seven other data sources relating to sea surface temperature and weather phenomena.

After the success with the Sahel prediction, Dr. Schmitt stumbled across a contest that challenged participants to develop a more accurate predictive model of seasonal weather in the western United States. He recruited his twin sons, who had some experience with programming and machine learning, and the trio developed a neural network trained to predict seasonal rainfall. The work showed that accounting for ocean salinity enhanced the predictive power of their model. They won the contest handily, then used the prize to launch a company, Salient Predictions, that leverages AI to aid farmers. Dr. Schmitt and his colleagues also demonstrated that their technique applies to the area that initially sparked his interest, the Midwest. Their model outperformed the standard forecasting methods by 92%, offering incredible potential for predicting seasons of heavy rainfall and drought. As with the traffic accident map, the model will not say exactly when to expect a downpour or flood. But it could give regional planners and farmers the opportunity to prepare for potentially damaging events. Indeed, Salient Predictions was awarded a Bill & Melinda Gates Foundation grant to help bring these capabilities to small farmers in East Africa.

• • •

What else might we do as we sharpen our foresight through AI? My favorite wizard could have used an AI-enhanced map to track or evade his foes, but parents of adventurous teens might use a similar tool, tuned to a standard smartphone location tracker, to not only see where their children are but guess where they might be headed. Similar to its successful application in meteorology, AI-powered foresight in business is revolutionizing the capacity to plan, adapt, and innovate in an increasingly complex world. Predictive analytics tools forecast supply chain demands and production needs in manufacturing, ensuring that companies do not produce too much or too little. Ride-sharing companies like Uber and Lyft use AI to project demand in different areas and at different times. This helps them optimize the allocation of drivers and reduce waiting times for customers, and also sets dynamic pricing. Alphabet uses AI to predict and optimize energy usage in their data centers. By analyzing factors like weather patterns and historical energy usage data, they can reduce costs and increase efficiency. AI-powered models like those developed by my colleagues Dina Katabi and Regina Barzilay and their teams can analyze patient data to predict the development of breast and lung cancer or the progression of chronic diseases like diabetes or Alzheimer's, helping patients and healthcare providers develop better long-term treatment plans. BlueDot, an infectious disease intelligence provider, used AI to successfully predict the 2019 Covid outbreak in Wuhan before it was officially recognized. In the insurance industry, Oscar Health uses AI to forecast claims and health expenses, allowing them to optimize pricing and improve risk management. The agriculture technology company Granular offers predictive analytics tools that provide foresight into the potential crop yield of different fields based on various factors, including weather conditions, soil health, and management practices. With tools such as these, my father, who so loves the grapes he grows, will know when to cover and bury the precious vines to protect them against frost.

When I consider all the possible uses, I cannot help but think again of Doctor Strange. This highly educated surgeon-turned-superhero has a number of wonderful powers. His ability to move rapidly around the world via wormhole is particularly appealing as a carbon-zero travel method. Yet that critical moment in which he envisions millions of possible outcomes of the Avengers' war against Thanos is perhaps the most exciting power of all, and it is one that may soon be available to all of us, in many areas of life. The predictive models will improve. The foresight of these tools will sharpen, and we will all soon be able to use them in concert with our own knowledge and insight to strengthen our ability to plan strategically for the fundamentally uncertain future.

6

Mastery

THE MOST POPULAR SPORT ON OUR PLANET IS FOOTball, or what most Americans refer to as soccer, but the runner-up might surprise you. Although it is largely limited to lawn games and physical education classes in the United States, badminton is enjoyed by hundreds of millions of people worldwide. The sport is particularly popular in Asia, and my lab is now working with Professor SeungJun Kim and his group at the Gwangju Institute of Science and Technology (GIST) in Korea to develop an AI-based analytics tool that will help everyone from beginners to world-class professionals improve their badminton technique.

The students started by filming players of different skill levels and then fed this data into an AI model, specifying whether the participants were beginner, intermediate, or expert. Basically, we tell the AI which players are good, bad, or in-between, and it maps the subtle connections within these groups. In the same way that a machine learning model trained on labeled images can eventually determine whether an unlabeled photo features a duck or a dog, the AI will learn to classify players it has never seen before and determine where they fit on the spectrum of skills.

Let's imagine I were to join a game in Korea and switch on the model. Obviously the AI would need some sensing—a camera to track my motions and the movement of the shuttlecock. First, it would determine my ability, then perform further analysis on my stroke and compute possible interventions or suggestions to help me improve. The model might tell me if the plane of my swing is off, for example, or whether I should accelerate more through my hit.

The exciting piece here is not the potential to revolutionize badminton coaching, but the fact that we could take this larger concept and apply it to just about any profession or hobby that involves physical skill. Golf instructors are already relying on AI-powered applications to help show their students subtle glitches in their swings. Surgeons can receive AI feedback as they perform operations such as suturing to help them improve their technique. And we do not need to confine these tools to physical activities, either. There are AI tutors demonstrating their value in education, helping students learn languages, scientific concepts, and more. Eventually they could help us learn anything, which reminds me of one of my favorite moments in the classic science fiction film *The Matrix*. The movie's hero, Neo, has only recently escaped the computer-simulated reality he's been trapped inside his whole life and has begun adapting to the very bleak real world. He is strapped into a chair and his brain is wired through a neural link to some presumably advanced computers. One of his new friends, the programmer Dozer, begins uploading various martial arts programs into Neo's mind. Eventually, Neo opens his eyes wide and declares, "I know Kung Fu."

If only we could all achieve mastery so quickly! I don't expect that AI will accelerate learning to that extent, but the tools we are developing today will expand our options for education, both as children and adults, allowing us to achieve proficiency and potentially even mastery at a faster pace and with less pain. These tools could allow more young people to benefit from the sort of personalized tutoring that has long been limited to the privileged few. Yet we have to be certain that we deploy these technologies thoughtfully, as the early examples in education have made evident.

When ChatGPT had only been available for a few months, one survey found that 60% of college students confessed to having already used the AI engine on more than half their assignments. A later survey revealed that 44% of teenagers planned to use the tool for the 2023–2024 school year. Yet I've also spoken with academics who state that these AI systems are not widely used in their institutions. These tools are certainly intelligent. Their performance isn't necessarily reliable, but they can excel; OpenAI has claimed that GPT-4 scored 1410 out of a possible 1600 points on the Scholastic Aptitude Test (SAT). As a professor, I'm naturally very concerned about how AI tools will change education. The impact on my own fields of computer science, AI, and robotics will likely be massive, and I can understand why a high school English teacher would feel despondent over the introduction and widespread use of an AI that can write a passable essay on just about any topic. Similarly, if a student leaves her coding to an AI model and reduces her work to a series of written prompts, will she really learn how to develop software or think computationally? I don't think so. Anyone who has aspirations in computer science needs to understand the basic languages involved and how to design programs both at a high level and at the level of the line.

Yet there may be some advantages to incorporating these tools in a strategic way. The teaching of writing offers a good example. When most people were panicking that the arrival of ChatGPT signaled the end of education, a high school English teacher named Kelly Gibson began experimenting with using the AI in a very different way. She developed a plan by which students would originate a thesis based on their reading of several texts. They enlisted ChatGPT to generate first drafts of their essays, then spent most of their time and mental effort revising and refining what the AI produced. Many teachers of young people will tell you that getting students to revise their work is an incredibly difficult step. This AI-powered exercise shifts the brunt of the effort away from the production phase, and keeps the focus on the initial idea generation and final refinement of the argument. In fact, the same shift emerged naturally in the writing productivity study detailed earlier. The professional writers in that experiment devoted more of their time to originating, editing, and revising their work when they made use of the large language model. They expended more effort thinking about their pieces at a high level.

The initial research in computer science education is equally intriguing. A group led by the computer scientist Paul Denny at the University of Auckland in New Zealand conducted a study with GitHub Copilot, the AI model that generates code from language prompts, and found that the program successfully solved half of a given set of typical introductory computer science class assignments. After a few tweaks, the AI successfully solved 60% of the remaining exercises. Such stories of AI passing tests with ease are rampant. Some are legitimate, others apocryphal, yet it's relatively straightforward to imagine a model that could perform at least as well as an average student on many tests and written exams, unless we fail them for plagiarism. But while AI can help students learn more effectively and engage with information in novel ways, the brain's ability to form long-term memories depends on the quality of the learning experience, the student's motivation, their overall cognitive health, and other factors. AI is a powerful tool to support learning, but it is not a substitute for the hard work, effort, and dedication required to develop long-lasting knowledge and critical thinking.

So what do we do? How do we move forward?

We can't lock AI out of the academy, especially if we hope to prepare our students for professional success in computer science. Many developers are already using these tools, so we need to teach our students how to use them, too. And yet our reasons do not need to be entirely practical or mercenary. There may be educational benefits, as suggested by another study looking at how AI tools impacted students in introductory courses. This one focused on sixty-nine middle- and high-school-aged students. The young people were given forty-five tasks related to coding in Python, a popular programming language. Some of the participants did the work manually, while others were allowed to use OpenAI Codex, another generative tool for writing code. The kids who used Codex had higher completion rates and overall scores, which is not surprising. Interestingly, though, they also performed well when asked to modify the resulting code without the help of AI, and they outperformed their peers on follow-up tests designed to measure how well they understood the code. One would assume that manual coding would lead to far better results in this area, yet it seems that working alongside an AI was more effective. Tools that teach this sort of thinking will be extremely valuable, and my colleagues Armando Solar-Lezama and Josh Tenenbaum are pushing this idea with DreamCoder, a system that learns to solve problems by writing computer programs. This tool has many potential uses; from an educational perspective, working with DreamCoder will help aspiring programmers think at a higher level of abstraction.

These are early studies, and the technologies are evolving rapidly. How we teach these skills will undoubtedly change. As the hype and excitement around generative AI tools swelled in early 2023, the computer scientists Daniel Zingaro and Leo Porter outlined a new approach to teaching programming that incorporated code-generating tools. In their vision, classes would place less emphasis on the actual coding, putting more mental effort into designing software, testing different variations of the AI-generated code, and reviewing and potentially debugging that code. Zingaro and Porter suggested that this shift away from teaching and learning syntax to higher-level computational thinking would produce more students with the skills to solve larger, more complex problems—in other words, students capable of producing working software after completing an introductory computer science course.

• • •

Forgive my focus on programming; the power of AI as an educational tool will reach far beyond computer science. Consider general youth education. One of the quiet tragedies of the field is that researchers have known for a very long time that personalized, one-on-one tutoring is an incredibly powerful teaching method. Yet it is also the least practical, at least as an institutional strategy. Attempting to provide a one-to-one teacher-to-student ratio in public or private schools would be impossible, so this approach is reserved for the wealthiest sectors of society. When we consider why tutoring is effective, though, the fit for AI becomes clear. Good tutors have strong knowledge of the subject. (Our

AI Librarians know everything.) They can sense and track a student's understanding of the material. (AI tools can monitor facial expressions and body language for signs of stress, anxiety, frustration.) And perhaps more importantly, a good tutor can adjust the pace or focus of the lesson to match the student's needs in the moment. (An AI tutor will never watch the clock.)

The positive impact of these virtual AI tutors is already becoming evident in academic research. In one study, the computer scientist and MIT CSAIL alumna Emma Brunskill led a Stanford University group developing and testing an intelligent math tutor that adapts its lessons in real time based on the participant's engagement, progress, and knowledge. The system has a goal—improving the student's test scores—and it adjusts how it progresses toward that goal based on the individual. The system also learns through experience, based on how effectively the new plan moves it toward its goal.

In their experiment, the researchers attempted to teach kids about the physical concept of volume through an interactive narrative. Children were tested prior to interacting with the tool for knowledge, math anxiety, and other attributes. They were given the chance to select from one of several predetermined friendly monsters to serve as the digital avatar of the tutor, and, as they worked through the volume-related exercise, this intelligent monster operated in a side panel on the screen. At first the AI asked questions about the student to forge the beginnings of a relationship and familiarize the kid with their tutor. Then the friendly AI monster interacted with the student throughout the exercise, offering generic or more detailed encouragement, direct hints, or guided questioning designed to help the student learn. Interestingly, the system had the greatest impact on students with lower pre-experiment scores. As with the ChatGPT experiment that helped less-skilled writers perform faster, the AI helped shrink the gap between highest- and lowest-performing students.

The math students had the opportunity to quit the game at any time, and yet they largely chose to continue. This points to one of the more promising elements of virtual tutors: AI tutors could be optimized to help students stay engaged. Emma and her colleagues noticed this in another experiment, in which they developed an intelligent chat interface called EnglishBot to help Chinese college students improve their conversational English. The participants spent twice as much time interacting with the virtual tutor, relative to working with standard automated techniques that ask students to listen to and repeat a phrase. Similarly, the language-teaching company Duolingo, which serves 50 million customers, leveraged GPT-4 to develop a feature that allows users to have conversations with AI agents, then receive feedback explaining what went wrong when they made a mistake. The vast corpus of text on which GPT-4 was trained allows this tutor to discuss a wide range of subjects, so users can approach the kinds of conversations they'd have in the real world.

Perhaps I am biased toward science fiction concepts, but I have always loved holograms and cannot help but imagine that virtual, three-dimensional tutors would only help in the effort to learn. One AI tutoring startup, Matrix Holograms, aims to deploy a virtual projection of a teacher to help drive engagement and learning progress. The technology will read the body language and facial expression of the student to track their emotional state and respond proactively if a specific question induces anxiety, for example.

Whether these particular academic projects and startup companies evolve into finished and widely available AI tutors I cannot say—we don't have a forecasting model for that yet—but this technology could be transformative. If we can develop effective virtual tutors at scale, and do so in a way that reduces the cost, we could make them widely available to children and adults who lack the means to pay a human tutor. We can start to think about extending a benefit that was once available only to the wealthiest members of society to everyone around the world.

This tutor would not need to be strictly educational, either. When I was a kid, I dreamed of having a magical book that could give me practical advice in the moment or teach me exactly what I needed to know in order to respond to a specific situation or solve a particular problem. Generally, AI excels at analyzing past data and patterns to suggest what to do in the moment. It's very good at *now*, and I can envision an AI-powered version of my magical book that could have acted like a counselor or a very smart imaginary friend. What if your diary was not only a confessional medium but an opportunity for interactive learning? This smart notebook could absorb your concerns and insecurities and then suggest strategies for overcoming them, helping you master one of the more complex systems in the universe—the self.

• • •

Mastery here is not just proficiency or skill in a given area, but a high level of command, understanding, and even artistry that allows for exceptional performance. If we expand our ambit, AI can be used to catalyze expertise in various businesses, too. AI can help individuals by identifying gaps in knowledge or skill, suggesting personalized learning pathways, providing real-time feedback, and enabling practice through simulations. Organizations could enlist AI to create personalized learning paths for upskilling and reskilling employees, based on their individual needs, goals, and current level. The AI-driven educational technology startup Degreed is already developing an AI platform that allows companies to personalize employee learning experiences. AI can incorporate gamification elements into training programs, making learning more engaging and motivating employees to master skills faster. Duolingo for Business allows companies to offer language training to their employees in a game-like format that encourages mastery. Driven young professionals aspiring to leadership roles could be trained through AI to become effective managers and even executives.

Hobbyists have long turned to the web to learn how to do everything from crocheting to playing the guitar to performing simple maintenance or repair work on a car or home appliance. Instead of constantly pausing, rewinding, and rewatching instructional videos, imagine an AI application that directs your actions in real time as you try something out yourself. You would not be limited to watching and mimicking a guitar teacher. An AI variation of that teacher could study your technique and offer feedback and tips. The same approach could be used when you are trying to fix your dishwasher. An AI butler could teach a young person how to properly knot a tie.

Personally, I would have loved to have an AI tutor when I was getting dressed for a party recently. My colleague Srini Devadas invited me to a Dussehra party, a Hindu holiday that celebrates the victory of Rama over the demon king who abducted his wife. I own a beautiful sari and wanted to wear it to the event as a sign of respect for my friend and the tradition. Unfortunately, properly tying a sari is not easy. I asked Google how to do it and was directed to an instructional video, but after watching and rewatching the clip and starting over numerous times, I had to give up because I was going to be late. I hurried to the party, where I had someone tie it for me, but a virtual coach would have been ideal. An AI application trained on the proper techniques, and a wide variety of saris and people, would have been able to guide me through the process step by step. In one sense, a sari instructor wouldn't be all that different from a virtual surgical instructor or badminton coach, from a technology standpoint. We'd feed the system the data it needs to extract patterns or common themes apparent in the work of experts, then show beginners and intermediates precisely where they differ or go astray. Within minutes, I'd be a master. The research community has been working on this idea in the form of virtual mirrors and virtual coaches for sports.

The impact of AI teachers will be felt both inside and outside of the classroom, as we start to envision a future in which it becomes much easier to learn anything quickly and advance from beginner to intermediate, or from average-level skill to mastery, in far less time than ever before. As an academic I'm dedicated to lifelong learning, but the excitement for me stretches beyond esoteric disciplines in mathematics, computer science, physics, and biology. I'd like to sharpen my skills as a chef, achieve fluency in more languages, learn a new musical instrument, improve my tennis game and swimming technique. Maybe I'll even follow Neo and learn Kung Fu.

7

Empathy

IN ONE OF THE FINAL SCENES OF THE CLASSIC NOVEL *Moby-Dick*, the white whale breaches magnificently, drawing the attention of the *Pequod*. The ship's captain, the vengeance-fueled Ahab, cries out to the whale that this will be its last such display. He orders his men into the whaleboats, then descends into one himself. As the boats race toward the leviathan, Moby Dick does not dash to the depths but turns and charges at the crew. The epic fight between this amazing creature and the harpooners and sailors is horrifically violent—and it does not end well for either of the main combatants.

While I realize that this would not be a very dramatic ending, I wonder what might have happened if Ahab and Moby Dick had been able to work out their differences through civilized communication. Would they have explained their grievances and agreed to a truce? Would the whale's explanation of his plight have inspired Ahab to transform himself into a pioneering conservationist? I realize this is unlikely, but I cannot help considering the possibility, as AI could soon allow humans and whales to communicate in a rudimentary way. I'm working with a large interdisciplinary team to develop tools to try to understand how sperm whales communicate—a project that might even lead us, one day in the future, to exchange messages with them, although it is not clear whether this would be beneficial to the whales. The true objective of the project is to promote conservation by decoding how sperm whales communicate; we hope to demonstrate and better understand the complexity of their individual and social intelligence and ultimately inspire empathy for these magnificent creatures.

Although empathy is not a traditional superpower along the lines of speed, strength, or flight, there is some precedence for its value in the comic book world. The twin-antennaed Mantis, a member of the Guardians of the Galaxy, has astounding empathetic abilities. At one point in the *Avengers* movie saga, she nearly saves the universe through emotional sensitivity. Empathy is an under-recognized superpower, which AI is already showing its capacity to enhance and support by improving and opening up new lines of communication.

The computational physicist Stephen Wolfram, creator of Mathematica and Wolfram Science, has likened ChatGPT to the first working telephone. Whereas that invention allowed two people to converse, however, a powerful language model can be a conversational interface between a person and . . . almost anything. We could use AI agents to talk to machines, software, even animals. Instead of simply enabling idle chatter, I'd like to see these tools deployed to enhance our empathy for the other intelligent creatures on our planet. Then again, we humans have room to improve how we communicate with one another, so perhaps we should start there.

• • •

One outcome of the AI boom is the increased use of more intelligent chatbots in the customer service space. I suspect that organizations won't simply hand their customer interactions to AI. Those relationships are too valuable; the competition would reinstitute humans to steal back their business. AI will likely be used to answer simply phrased, basic questions before passing the more complex ones to a human agent. I hope this is the case; there's nothing worse than being caught in an endless interaction loop with a mediocre chatbot. Yet early work has found that these more intelligent tools are surprisingly effective—and that they have the unexpected benefit of promoting empathy.

The results of a study of 5,179 customer support agents working for a Fortune 500 software firm largely focused on speed, as the analysis found that use of an AI-based conversational assistant increased productivity—or how many issues the agents resolved hourly—by 14% on average. The impact on less experienced agents was even more significant. The AI made them better at their jobs, allowing them to perform at a level closer to their more skilled counterparts. The researchers hypothesize that this happened because the AI studied the techniques of the effective agents and replicated them; those agents had less to gain because the technology was borrowing their tricks. Yet there was also a surprising result: customers treated the human agents better when they did interact and were less likely to ask for a manager. The AI actually strengthened subsequent human interactions.

There is evidence that these tools could help doctors as well. In another study, researchers asked patients to compare different responses to standard medical questions. A large language model answered the queries in some instances, and in others actual physicians responded. On average, the patients preferred the AI-generated answers. They actually judged them to be more empathetic than those written by their fellow humans. Perhaps doctors who struggle with their bedside manner could enlist these models to enhance the empathetic quotient of their interactions. In their book *The AI Revolution in Medicine and Beyond*, Peter Lee, Zach Kohane, and Carey Goldberg discuss how ChatGPT is capable of providing effective advice to doctors on how to communicate with patients in a more compassionate way. The technology itself is not empathetic. Nor does it understand this concept. But it can simulate empathy and effectively teach us by example.

If AI can help doctors become more empathetic in their communications, then it could be an incredibly powerful tool for teaching children this quality, too. Normally, children begin to learn empathy at four years of age. Parents and caregivers are essential in this teaching process, as they model empathetic behaviors and actions for kids, by caring for others and for the children themselves. I wonder if we could recruit toys to help in this effort. I do not think dolls and action figures should necessarily come to life in the mode of the movie *Toy Story* and share their own fears, dreams, and insecurities—and we are very, very far away from developing a robot with anywhere near the capabilities of the AI-empowered doll in the 2022 horror film *M3GAN*—yet we could build relatively simple artificially intelligent toys that teach children the value and importance of caring for others. Barbie would be a perfect candidate. Although the long history of this doll has involved some controversy, I'm thrilled to see her current burst of popularity, as I have always found her to be inspiring. Never mind the occasional marketing missteps: Barbie has been a doctor, an astronaut, an engineer, and an executive. In the 2023 movie *Barbie*, she is a president and a Nobel Prize–winning physicist.

Let's say we were to endow a Barbie doll with some AI-powered interactive capabilities, a microphone, and a speaker, and design a strong security strategy that protected the privacy and personal data of the child who plays with the toy. This highly intelligent Barbie could teach children all manner of lessons using language that they would understand—a physicist Barbie could explain why the sky is blue, for example—and model empathetic behaviors. Maybe this AI Barbie could encourage the child to pet the neglected family dog, for example, or share stories of how her character has helped others in the past.

* * *

Empathy stems from understanding another living being and putting yourself in their shoes, and there are tremendous opportunities to use AI to strengthen empathy between humans by building our knowledge and understanding of one another. This has long been the role of books and movies, and my MIT colleague D. Fox Harrell demonstrated the possibilities in a virtual reality film designed with the photographer Karim Ben Khelifa. After years of meeting and photographing combatants on both sides of violent conflicts, Khelifa wanted to find a better way to help people see these fighters as humans.

His project with Fox, "The Enemy," featured two subjects, one an Israeli soldier and the other a member of the Popular Front for the Liberation of Palestine. Their work was virtual, interactive fiction that explored the themes of identity, empathy, and conflict resolution through the use of avatars and a unique storytelling structure. Participants interacted with the avatars by engaging in various conversations and making choices that impacted the narrative and the relationships between characters. To create the avatars, Khelifa asked the Israeli and the Palestinian the same six questions, including "Who's your enemy and why?" and filmed their responses. Then, using three-dimensional scans, Fox, Khelifa, and their collaborators crafted realistic avatars of each man and created a virtual reality film in which the digital versions of the combatants expressed their answers. As the viewer, you listened to each man's answers and felt as if you were in the space with them and they were addressing you directly. The game tracked the emotional states of your avatar and the other characters, and encouraged participants to resolve conflicts through empathy and understanding, rather than violence or confrontation, and the result was a far more engrossing experience than the standard documentary.

This technology is by no means a cure for the conflict, but it provoked an incredible sense of empathy in the viewer/player. It could be a first step toward using technology to help people gain a deeper understanding of their purported enemies as human beings. In his current research, Fox continues to look at how AI and gamification can be used to create a deeper understanding among people by designing virtual role-playing game scenarios centered around xenophobia and bullying, using an interactive narrative powered by an AI engine.

Another critical technology movement on this front is AI emotion recognition, an active research field that seeks to identify and track human emotions by analyzing cues like facial expressions, voice tone, and body language. My MIT colleague Rosalind Picard is a pioneer in this emerging domain. The process begins with collecting vast amounts of data, such as images or videos of people expressing different emotions, or audio recordings of speech with various emotional tones. The AI system then extracts key features from this data—a smile or raised eyebrows from an image, or variations in pitch and volume from an audio clip. These features are used to train machine learning algorithms that learn to associate them with specific emotions. Once trained, the AI system can classify new data by emotion. Given a new image, video, or audio recording, it identifies the relevant features, analyzes them with the trained algorithms, and assigns an emotion to the input. For instance, a smile and raised eyebrows might be categorized as happiness.

The technology is still evolving. The accuracy and reliability of these systems can vary, they can be prone to bias, and there are important ethical considerations regarding privacy and potential misuse. Yet if we work out these flaws, technological tools that allow us to peek beneath the surface of someone's facial expressions and body language or pick up subtle vocal nuances that help us understand what that person might really be feeling could be useful in many areas of life. I would have enjoyed such tools when my daughters were teenagers, for example, and the gulf between what they said and what they meant was probably wider than I understood at times.

While knowledge and understanding are critical, the primary vehicle for promoting empathy remains communication. I'd like to see AI bring about an end to the language divide between people. We're not there yet, but we are getting closer. If you and I were to chat, you could use the wrong word now and then, or speak in grammatically incorrect sentences, and I'd still understand what you mean. We'd still be able to connect. So, with that in mind, what if we could instantly but imperfectly translate real, live conversations? In the classic work of science fiction *The Hitchhiker's Guide to the Galaxy*, the characters stick creatures called Babel Fish in their ears and are instantly able to speak to aliens in different languages. I'm not sure I'd put a fish in my ear, but I don't want to rely on speaking or typing into a smartphone app, either. I have tried; the process is slow and awkward.

Instead, I'd like to have naturally paced, culturally appropriate chats with people in Tokyo or Seoul or Spain in their own language, to connect with them on a deeper level. We have the data. What we need is enough local processing power to parse one person's utterance, quickly translate it for the listener, then transform the spoken response into the other language. While I have done extensive work building intelligent robotic fish, I do not think we are ready yet to build one small and capable enough to facilitate instant translation inside your ear. Still, this sort of communication through instant, imperfect spoken translation is certainly possible. In some sense, it isn't even that bold an idea. My students are far more audacious. When I asked my research group what sort of capabilities they'd like to enhance through AI in ten years, they answered that they wanted to be able to communicate with one another silently, using only their thoughts.

An equally audacious possibility is the use of AI to enhance our communication with our companions in the animal world. Our diverse team of experts studying how we might use AI to understand the language of sperm whales was inspired in part by the work of the biologist Roger Payne, who sought to promote global conservation and animal welfare. The more we can demonstrate that sperm whales and other creatures are intelligent, social creatures with rich mental lives, the more likely our own species will back efforts to preserve and protect them. This is an immensely challenging and immensely fascinating project. We know that animals signal each other. But we don't know whether we can compare this to human communication. In other words, we aren't certain that we can map what we understand about how humans communicate to the barking of dogs or the chirping of birds or the distinctive utterances of whales.

Our group chose to focus on sperm whales because they exhibit impressive cognitive abilities and social behavior, and they emit audible clicks that could harbor some communicative intent or meaning. But how do we make sense of these noises? Large language models and diffusion tools have been successful in part because of the extensive database that is the Internet. Any successful current AI tool needs a massive dataset, so we're now beginning to compile and build one for sperm whales. This isn't merely a matter of recording their sounds. In animal communication, context matters. The vervet monkey, for example, issues an alarm call when it spots a predator, signaling others in its group to hide. Merely listening to this call would not be enough to understand it; we'd need to capture data that covers the entire scene and all the individuals involved. For whales, we need aerial and underwater drones, sensors on the creatures themselves, and acoustic recorders that can pick up distant sounds. This won't be easy work. The marine world is cacophonous. Most of the 126 marine mammal species are known to make noise, along with 1,000 species of fish, and researchers today are collecting more underwater sounds than ever before. One group has called for building an open global research library of all underwater sounds.

Even if we could isolate enough distinct sperm whale sounds, our models could not be confined to studying just this one type of data. They'd be looking for patterns and relationships between acoustic information, video, geopositioning information, and much more. We'll need to know what the whales are doing when they make each noise. And once we have models that can find patterns in this data, we'll need to develop linguistic models that can translate these patterns back into words, so we can communicate with meaning and intent. This might take some time, but as a group we believe it's achievable, and we're not the only scientists working on such challenges.

There is an opportunity to enhance empathy between more creatures, people, and systems that is truly exciting and eye-opening, especially when these possibilities are considered alongside the other applications we have discussed thus far. It's also important to remember that I'm not talking about one AI system here, but many specific and finely tuned solutions. As these systems and the startups or organizations behind them proliferate, though, we can take some comfort in the knowledge that today there are four fundamental approaches at work in the vast majority of these systems. These are approaches that have been studied and researched for many years. We know how they work, and the more we expand this fundamental understanding beyond experts and academics, the better the chance we have of ensuring these technologies are deployed in a safe, responsible, and ethical way that benefits the largest number of people possible.

So let's get into the fundamentals.

PART TWO
Fundamentals

RECOGNIZING THE CAPABILITIES OF AI IS ONLY THE beginning. The true challenge—and opportunity—lies in operationalizing them, translating theories and ideas into actionable, scalable systems. In Part Two, we delve deeper into how models are constructed, the principles guiding their formation, and the pivotal role these fundamental design decisions play in turning AI into truly useful tools. The successful integration of AI into our lives, workplaces, and society hinges on a basic understanding of these principles, not only by experts but also by those without a technical background. Deep technical knowledge is essential for anyone developing AI systems. However, it's useful for everybody to gain and cultivate an understanding of how AI works, its capabilities and limitations, and when and how to effectively leverage its strengths.

First, let's review the distinctions. The terms "artificial intelligence" and "machine learning" are frequently used interchangeably, but they are distinct academic technical fields and serve distinct roles. AI's overarching goal is to craft systems capable of executing tasks that typically require human intelligence. This can range from recognizing patterns to more complex functions like playing games, moving in the world, perceiving the world, reasoning, problem-solving, understanding natural language, and even learning. Although many AI systems can learn from data to establish and improve their performance, this is not a prerequisite. An AI chess program, for example, could be programmed with a predefined set of rules and play the game at a very high level without requiring data from past games of chess—in other words, without actually learning from its past mistakes. Some AI systems might execute a specific task perfectly every time, based on their programming.

You might also hear of "Narrow" or "Weak" AI—these are systems that excel at a specific task, such as image recognition or voice response. On the other end of the spectrum lies the conceptual "General" or "Strong" AI. This still-theoretical brand of AI would generate human-like cognition, making it versatile and adaptable across any task or field. These are the sentient machines we often see in movies. We are far away from conscious or sentient AI systems from a technical standpoint. Sentience refers to the capacity for subjective experiences and feelings, like pain or joy, while consciousness denotes a state of being aware of and able to think about one's existence, sensations, and thoughts. These complex, inherently human attributes are beyond the scope of current AI, which operates on predefined algorithms and lacks the ability to experience subjective sensations or self-awareness. The emergence of sentience or consciousness in AI systems, as of now, is a theoretical concept more rooted in the realms of philosophy and speculative science than in practical computer science. As of the writing of this book, no such systems have been developed, and there is no obvious pathway to building them.

Whereas AI has a broad scope, encompassing any system that enables technology to mimic human intelligence, machine learning focuses specifically on the development of algorithms that can learn from and make predictions or decisions based on data. You don't need to program a machine learning solution explicitly. Given quality data and carefully selected algorithms, a machine learning system can figure out how to accomplish a task and improve its performance over time. If you show a machine learning model numerous pictures of cars, it will eventually recognize and categorize new images of cars on its own. This iterative approach to improvement is driven by an ongoing feedback loop in training—the model makes a prediction, measures the accuracy of its prediction against the actual outcome using the knowledge embedded in the training data (for example, human-affixed labels that specify that the subject of a photo is a car), adjusts its parameters, and then tries again.

Machine learning is deeply intertwined with statistics. For instance, to recognize an object in a picture, a machine learning model might use statistical methods to weigh the importance of each pixel and how it correlates with the identified object. The feedback loops allow the models to refine their predictions and actions, and these too are embedded with techniques from statistics. These techniques enable the models to evaluate the probability and likelihood of various decisions and resulting outcomes, facilitating continuous improvement in performance. All of these elements combine to enable the model to make highly informed guesses, check its work, and sharpen its approach.

AI and machine learning are sometimes interchanged because machine learning is often embedded within or operating in the service of AI solutions, from Netflix's recommendation systems to Siri's voice recognition. Essentially, while AI sets the stage for machines to mimic human intelligence in various ways, machine learning provides the tools that allow us to enhance an AI model's performance through experience. Think of this in terms of how they may be used by robots. While AI and machine learning concern themselves primarily with data, algorithms, and decision-making, robotics brings these concepts into the tangible realm, giving AI and machine learning a physical presence. Robotics, the subject of our book *The Heart and the Chip*, deals with the design, construction, operation, and use of machines that can perform physical tasks autonomously. The brains of these robots, their decision-making prowess, often depend on AI. For instance, a robot navigating a room might use AI to interpret its environment and decide where to go, and machine learning to adapt after making mistakes, improving its navigation over time. Thus, AI provides the cognitive capabilities—the thinking—for the robot; machine learning ensures that this thinking improves with experience; and robotics offers the mechanical means to act on that thinking. It's a harmonious blend: robotics gives AI and machine learning a way to interact with and learn from the physical world, while they provide robots with the intelligence to operate in diverse and unpredictable environments.

There are many different algorithms, models, and architectures for machine learning and AI solutions, but the primary solutions can be categorized into predictive AI, generative AI, reinforcement learning, and decision-making, or combinations of these methods.

8

Predicting

PREDICTIVE AI RELIES ON ALGORITHMS AND MODELS TO forecast the likelihood of future events or outcomes based on historical and current data. These solutions depend on machine learning systems that recognize and classify patterns in data, then make predictions based on those patterns. Integral to these systems is their ability to not only recognize but also discriminate between patterns corresponding to different categories in the data, enhancing the accuracy of their predictions. Think of this family of models as specializing in making highly informed estimations. These predictions could be about some future event, or they might be an inference about something in the moment, like classifying the subject of an image as a car. So, how does Predictive AI work? How do we design these systems?

The first step is to acquire a high-quality, unbiased dataset. This data should be accurate, relevant, extensive, and devoid of mistakes. For instance, an image labeled as a parrot should depict a parrot, not a bear. This data may be drawn from various sources, including databases, sensors, user inputs, and more. Sometimes there will be errors, so, after collecting data, it is important to cleanse it by rectifying or eliminating any inaccuracies or inconsistencies. Following this, we reformat the data to align with our model's requirements.

Next, we choose or design a specific architecture for the model and train it using the appropriate algorithm or combination of algorithms. This architecture—essentially, the framework or structure of the model—dictates how it will process and interpret the data. (One of the better-known architectures is the neural network, discussed below.) We divide the data into two subsets, using the first to train the model and the second to evaluate its effectiveness after training. The training process is iterative and involves continually adjusting the model's parameters to minimize errors and improve accuracy. Following training, we conduct thorough testing with the second data subset to validate the model's performance, ensuring it can accurately generalize its predictions or decisions to new, unseen data. This gives us some idea of how it will perform in the real world. All the while, we use various tools and techniques to enhance accuracy, mitigate biases, and generally optimize our model's performance. This development process is iterative as well, with models frequently being fine-tuned as more data becomes available, environmental factors evolve, or new errors and flaws are spotted. The models need to be continuously monitored to ensure they remain accurate and relevant, and that they do not produce biased or discriminatory suggestions. If the data relevant to the application changes significantly, then the model might need to be retrained. A housing price prediction model that doesn't incorporate and adapt in response to a recent crash in the market would not be very useful.

Let's look at one of the more popular and effective Predictive AI models, the neural network. This architecture consists of nodes called artificial neurons or perceptrons. Each artificial neuron is connected by edges, in much the same way that the neurons in our brain are linked by synapses. Multiple edges can feed into a single node, and that node can then connect to additional nodes through more edges. Each edge (or connection) between two of these artificial neurons has a weight, which determines the strength and sign (i.e., the signal should be positively or negatively reinforced) of the connection. Generally, a neuron receives inputs through the edges, performs some computation, then passes the results on to the next neuron(s) to which it is connected. The nodes also have a bias—this is a mathematical, not an ethical term—that can change the computation inside.

Adding more artificial neurons to a network increases its ability to represent more complex functions and process larger amounts of data. However, after a certain point, this addition yields diminishing returns in terms of performance improvement. The larger models generally require more training data to avoid what we call overfitting, which occurs when a model is so intricately adjusted to its training data that it fails to perform well on new or unfamiliar data. To train these models effectively, so that they can capture the sort of hierarchical patterns in the data that will prevent overfitting, AI researchers organize them into layered structures in neural networks. This layered arrangement—an added dimension to the neural network architecture—allows the model to process information at various levels of complexity and abstraction. Increasing the number of layers in a neural network enhances its potential capacity to represent functions. Networks with many layers are termed deep neural networks.

You might have heard the word "parameter" thrown around

in discussions of various AI models. In the context of neural networks, parameters include weights (associated with the strength of the connections between edges) and biases (associated with the computation happening inside the nodes). When people say that the foundational model GPT-4 has a trillion parameters, it signifies there are about one trillion weights and biases in the model. A model that has 218 billion parameters has 218 billion weights and biases. A useful, albeit simplified, comparison is with the human brain, which boasts about 100 trillion synapses or connections between neurons. However, it's essential to remember that the nature and function of AI parameters and human synapses are vastly different.

Anyway, let's get back to the layers.

Organizing the networks by layer allows for the emergence of different functions or tasks within each layer, enhancing the network's efficiency and precision in processing data. In simple terms, you can think of the first layer's role as receiving the input data. The final layer produces an output—the prediction or guess. Between the input and output layers, there are hidden layers that analyze and transform the data, capturing various levels of abstraction.

So how do we train and improve these models? In a deep network, we start by feeding the training data to the input layer. The artificial neurons in the network perform simple calculations based on the data they receive from preceding nodes, facilitated by the weights on the edges, and then relay their results onward. The output of one layer becomes the input for the next layer. Eventually, after all the nodes across all the layers perform their calculations and pass along their results, the final output layer produces a result or prediction.

In the training phase, we generally know the correct answer.

By comparing the model's prediction to this known answer, we assess the deviation, then adjust the network's weights to better align the predictions with true outcomes.

This is when the magic happens. There are a few different techniques for improving performance in deep networks, but one of the key innovations of the last few decades is the backpropagation algorithm (often abbreviated as backprop). When data is first fed into the neural network (what we call the forward pass), the network makes a prediction based on its current settings. We then measure how far this prediction is from the desired outcome. This gap is the error. The essence of backprop is to figure out which parts of the network contributed most to this error, much like retracing your steps to identify a mistake. The backprop algorithm starts from that final output layer and moves backward through the network, reviewing what happened where, and figuring out which parts of the network were likely responsible for the error. Then the network tweaks its settings, with larger contributors to the error receiving bigger adjustments. The model computes the changes using optimization methods like gradient descent* in order to adjust the network's weights to reduce this error. This cycle of prediction, error measurement, and adjustment is repeated many, many times during the model's training. Basically, we hone its accuracy through systematic adjustment of weights to drive down the error.

Imagine I'm trying to teach a robot to hit a half-court shot in basketball. The goal is for the robot to sink as many shots as possible. First, the robot shoots. This is the equivalent of what we

* Gradient descent is a proven and widely used algorithm that iteratively adjusts the network's weights in a way that reduces the error between the predicted and actual outcomes, thereby improving a model's accuracy over time.

call the forward pass, or the first time we push data through the network. Think of each shot as analogous to a single iteration of training, or the robot's hypothesis about the right way to score.

The AI system knows the difference between a make and a miss—in this example, through direct observation. The robot is supposed to see the ball go through the hoop, and if it misses, the model evaluates why. Was the shot too strong? Too weak? Too far to the left or right? If the shot was too strong, the system will subtly adjust its technique to apply a little less force, and try again. As the robot keeps shooting, it continues adjusting, getting better and better, learning from each successful and unsuccessful attempt. Over time, the robot amasses a refined understanding of the actions and conditions leading to a successful basket. This is how neural networks optimize over multiple iterations. A predictive AI system will evaluate how far its output deviates from its goal—the equivalent of a made basket—and tweak the network until those deviations get smaller and smaller. Eventually, the robot starts making every shot and the network starts producing more and more accurate predictions. While the ultimate objective is for the robot to consistently make shots and for the AI system to provide highly accurate predictions, perfection is elusive. Still, with adequate training, both can achieve remarkable proficiency.

There are different ways to interconnect the neurons or organize the layers within a deep network. These design decisions result in distinct architectures tailored for specific tasks, such as image classification, language translation, gameplay, shooting basketballs, or predicting stock market fluctuations. Analogously, think of the internal combustion engine. There is one foundational design for this engine, but there are many variations that can be optimized for efficiency, power, or speed. Similarly, the design of a neural network can be fine-tuned to serve a specific objective.

This isn't entirely straightforward, however. Deciding on the architecture is a blend of art, science, and patience. Achieving optimal results requires experts who grasp the nature of the data, the specific task at hand, the intricacies of the domain or industry, and the computational challenges involved. An architecture based on convolutional layers* might be best for image data, for example, because this approach is better at identifying spatial hierarchies and patterns such as edges and shapes within pictures. But a recurrent neural network (RNN) will be better for sequential data like stock prices, sentences, or video frames. Unlike the neural networks that process data one at a time independently and capture the results in the form of weights, RNNs have a memory of previous inputs. This memory is built through loops within the network that allow information to persist. As data flows through the network, each node in an RNN can use information from previous steps in the sequence to influence its output and adjust its responses based on this accumulated knowledge. In a sense this makes the nodes more effective than those in a standard neural network, and this characteristic is important for tasks with sequential or time-series data, such as the readings of a thermometer at various points throughout the day, because it allows the network to recognize that these values have changed over time.

Optimizing predictive models involves a good deal of experimentation. To understand the errors a model makes, practitioners

* A convolutional layer in a neural network employs filters or computational elements called kernels that slide over the input data, such as an image, to perform mathematical operations and sum up the results. This process allows the network to capture spatial hierarchies and patterns, like edges and textures, making convolutional layers particularly effective for processing visual information.

sometimes employ a confusion matrix, which categorizes predictions as true positives, false positives, false negatives, and true negatives. Analyzing this matrix offers insights that guide refinements to the model, such as employing different algorithms, tweaking the architecture, or expanding the training data.

The end goal, however, is the same. We want to develop a model that can make accurate predictions when given new, unfamiliar data outside of the training and validation subsets. By measuring its performance, understanding its errors, and adjusting accordingly, we start to gain confidence in its ability to make predictions, which we can use, in combination with our human knowledge and judgment, to make better, more informed decisions.

Hopefully I haven't lost you in the confusion matrix; to understand predictive AI, you need to appreciate its intricacies. These models glean insights from historical experiences or data, as humans do. However, their predictive prowess is bounded by the quality and diversity of this data. Biased data will usher in skewed outcomes. Striking the perfect balance between over-complexity and over-simplicity in model design ensures we avoid pitfalls like overfitting or underfitting.* We also have to recognize that these models learn in ways distinct from human cognition. While humans imbue objects and concepts with layers of meaning, machine learning models remain fundamentally driven by pattern recognition. For instance, when a model identifies a cup, it's merely because it has seen similar pixel patterns labeled by people as a cup in its training data. But it doesn't grasp the semantics

* Overfitting occurs when a model is too closely tailored to the training data, capturing noise and anomalies as if they were significant patterns, which leads to poor performance on new, unseen data. Conversely, underfitting happens when a model is too simple to capture the underlying patterns in the data, resulting in inadequate performance on both training and new data.

or inherent cup-ness of the object; it doesn't understand that cups can hold liquids, that the liquid may spill out if the cup is held in the wrong orientation, or that one of these containers might be ceramic and another paper. It's just pattern-matching the pixels.

This is not to diminish the value of pattern recognition, which enables accurate predictions and classifications even in vast data landscapes beyond human scale. Instead, it's a reminder of the models' lack of semantic depth. While their predictions can be immensely powerful, they often shine brightest when paired with human intuition and judgment, melding cold computation with warm comprehension. As AI continues to shape myriad sectors, from healthcare to criminal justice, the symbiosis of machine learning and human understanding is vital. Ethical and informed human decision-making must remain our north star, and embracing the strengths of AI while understanding its limitations will ensure we harness its potential responsibly and effectively.

9

Generating

THE BRAND OF AI THAT HAS SPARKED THE MOST INTEREST in recent years is different in that it actually produces new material. Instead of classifying data or predicting outcomes, generative models analyze the makeup of documents, images, videos, and other forms of data, then learn how to produce new variations, or new samples that are similar to but different from their training data. In the world of AI, there's an intriguing dichotomy between predictive and generative models. When we think of predictive AI, we're often talking about many-to-one functions: a multitude of distinct inputs lead to the same predicted category or outcome. For example, in image classification, various images of dogs, each with a different background, position, and lighting, could all map to one prediction label: dog. The AI system is recognizing the commonality among diverse inputs and categorizing them under a unified concept. On the other hand, generative AI operates in a one-to-many fashion: starting from a singular concept or idea, the model could generate a multitude of outputs. A generative model can create diverse, brand-new images of dogs from a single prompt such as "dog." Here, we're not narrowing down but expanding, constructing a myriad of details from a singular input. While predictive models aim to assign a label or value to an existing piece of data, generative models use learned concepts to dream up entirely new original data.

As of today, there are a few different generative AI algorithms. I will briefly introduce four approaches to generating data, and then follow with a more in-depth discussion of the algorithms that power today's most popular generative AI tools.

Convincing deepfake images, videos, and audio clips are often produced by Generative Adversarial Networks, or GANs—what Refik Anadol used for his exhibition. These consist of two models, a generator and a discriminator, trained together at the same time. The generator tries to produce fake data to fool the discriminator, while the discriminator tries to distinguish between real and fake data. At its core, the generator is like a skilled forger trying to create a piece of art that looks like a famous masterpiece. Initially, this forger doesn't know how to make convincing art. It starts by making random brush strokes on a canvas. Over time, as it receives feedback on its work, it refines its technique. Its goal is to produce artwork so compelling that even an art expert would have difficulty distinguishing it from a genuine piece.

In more technical terms, the generator takes in random noise (a random set of numbers) as input and produces data (like an image) as output. The brush strokes are analogous to the data the generator produces, trying to resemble the real data. The discriminator acts like the art expert or detective. It examines the artwork and decides whether it's an original or a creation of the forger. Each time the discriminator reviews a piece, it provides feedback to the generator, helping it to improve. If the discriminator gets fooled and thinks the fake data is real, the generator has done an excellent job.

Another model, the Variational Autoencoder (VAE), encodes the data into what we call a latent space—a variation that consists of only the most important features in the dataset—then decodes from this space to generate new data. Imagine you want to describe the appearance of a tree to someone who's never seen one. You might describe its narrow cylindrical trunk and leaves or the shape of its canopy, distilling the complex organism into these features. This simplification is akin to creating a latent space representation in AI, where complex data is condensed into its core characteristics. Now, if the person to whom you're explaining the tree is a skilled artist, they might be able to draw various trees based on this description, but the trees will look even better if more features are added to the explanation. Similarly, VAEs find a condensed, essential representation of the data in a latent space and then use this foundation to generate new, accurate data instances. Enhancing the latent representation of VAEs with more features, such as adding details about the color and texture of the bark to the description of the tree, leads to the generation of more refined and diverse outputs.

Transformer-based generative models such as the various GPT models can generate coherent paragraphs of text, essays, and papers. At their core, transformers operate by focusing on different parts of the input text to understand context and relationships between words, which allows them to generate coherent sentences by predicting one word at a time. Stable Diffusion models are deep neural networks that generate new images or objects. Imagine them as artists who create their works by tuning knobs; each knob decides different aspects of the picture, such as its color, shape, or theme. This is actually a complex mathematical process that helps the model randomly decide on various characteristics from the vast multitude of possible choices, leading to unique and diverse creations every time. The powerful generative AI solutions that have emerged in recent years owe much of their capability to breakthrough advances like the transformer architecture and emerging techniques like the Stable Diffusion algorithm. But let's start with the former, which is the engine behind ChatGPT and other large language models.

TRANSFORMERS

Think of a transformer as a very complicated machine with lots of knobs that can be adjusted to make the machine work better. In AI terms, these knobs are the parameters of the model. A large transformer model, like OpenAI's GPT series, can have billions of them.

Transformers are trained on text by masking words in sentences and trying to guess the missing words over and over again until the model guesses the words correctly across all sentences in its training data. Technically, the transformer is a machine learning model designed with a deep learning architecture to generate human language. It is capable of capturing complex patterns, nuances, and structures in language, generating text, answering questions, translating, summarizing, and more. Transformers work by a process called next token prediction. They revolutionized the AI field by using a mechanism called self-attention, which allows the model to weigh the significance of different words in a sequence relative to one another.

Now they work with video, too, but imagine we're trying to build one of these large language models.

First, we need to plan out an extensive, slow, and very expensive training period. The models need to train on trillions of words of text spread across website pages, books, digital news articles, blogs, and academic papers on the Internet.

Each word the model encounters during its training is assigned numerical coordinates within a virtual space of very high dimension known as an embedding. Think of the standard *x*, *y* graph in elementary school mathematics. Given a value for *x* and another for *y*, most children can find a point on a two-dimensional graph. When we add a *z* axis, the graph becomes a little more difficult to work with on the page, but we can all wrap our heads around three-dimensional spaces. Visualizing where the point resides is relatively easy. Now imagine we keep adding dimensions. Four, five, six, . . . a hundred . . . and so on. A foundational model* for one of these text-generating AI engines might have more than ten thousand dimensions. I won't ask you to picture this; it is beyond human comprehension.

During the training phase, words and word fragments are assigned various coordinates within this higher-dimensional space. When the model sees a word next to a different word in a sentence, or when the word is situated in a different place within a sentence relative to the first few times the model came across that word, then the word's coordinates change. New details are added as the *x* and *y* become attached to a *z* or an *a*, *b*, or *c*. Within the higher-dimensional space, the word shifts. Eventually, as the

* Foundational models are large-scale machine learning models pre-trained on vast amounts of data to understand general patterns and structures in that data. Once this base understanding is established, these models can be fine-tuned on specific tasks or narrower datasets. Fine-tuning refines the foundation model's capabilities, tailoring it to deliver specialized and more accurate results for a particular application.

model trains on enough documents and lines of text, words begin to settle into certain areas. Patterns emerge. Verbs cluster near one another. Names tend to end up near one another as well.

Why all this hyper-dimensional weirdness? Transformers are computer programs. They don't understand words. They understand numbers, and the more specific, the better, so the words and fragments in each sentence are converted into smaller chunks called tokens. The word "barking" becomes two different tokens, "bark" and "ing."* Each of these is associated with a string of numbers that represent their position in the hyper-dimensional word space. By situating all of the words and fragments in the English language within this space, we can predict, given a sequence of words or a sentence, the next word or a missing word with startling accuracy.

The transformer architecture became the foundation for numerous state-of-the-art models, including BERT, GPT, T5, LLaMA, and many others. The original model, developed by a team of researchers at Google and the University of Toronto, centered on a breakthrough idea. Instead of analyzing sentences sequentially, or one word at a time from beginning to end, their so-called self-attention mechanism looks at entire sentences at once, then weights the parts of the sentences differently.

How does that work?

Consider the following sentence:

THE SPOTTED DOG IS BARKING AT THE DOOR.

Imagine you're trying to understand the sentence, and you're

* In transformer models, splitting complex words into simpler tokens—for instance, dividing "barking" into "bark" and "ing," or "barked" into "bark" and "ed"—helps the model recognize word roots and grammatical structures, which enhances its ability to generate text that is more grammatically precise and contextually appropriate.

focusing on the word "barking." To get a clear idea of what's happening, you'd pay more attention to "dog" because it's the subject doing the barking. (Also, "barking" would actually consist of two tokens, "bark" and "ing," but let's disregard that for the sake of simplicity.) You might notice "spotted," which would give you a sense of detail, but it's not as crucial for understanding the main action. Instead, you might devote more attention to "door" because that's where the dog is barking.

The self-attention mechanism is applied to every word in the sentence. The mechanism focuses on a word, then relates it to all the other words in the sentence. Take the word "barking." With that as the focus, the mechanism computes scores on how much attention it should give to every word in the sentence, including "barking." In this case, "dog" and "door" might get higher scores because they're more relevant to the action of barking. The model doesn't understand or know that dogs bark; these words are closer to each other in that hyper-dimensional space because they are often close together in the text the model has trained on. Words like "spotted" or "the" might get lower scores because, while they're part of the sentence, they're less directly relevant to the action of barking. The attention mechanism helps the model weigh the relevance of each word in the context of "barking" and performs the same analysis on each of the other words.

After computing these attention scores, the transformer generates a weighted summary of all the words in the sentence, with each word's influence on the larger sentence determined by its score. For the word "barking," this summary will be more heavily influenced by "dog" and "door" due to their higher relevance scores; the process determines that they are more closely connected to "barking" in the sentence. In a sense, it is like a magic mathematical magnifying glass that amplifies certain words and minimizes others, and it is this ability to determine what is important within a sentence that makes self-attention so powerful. Instead of looking at words in isolation or within a fixed window, self-attention dynamically analyzes the entire sentence at once, studying how each word relates to all the other words. This mechanism helps the models understand more complex relations and dependencies in sentences. It's the reason ChatGPT is able to generate readable and logically organized text so quickly. The transformer attention mechanism also allows the simultaneous processing of all tokens, or fragments of words, which is why it can so efficiently capture relationships and dependencies among these tokens, regardless of their distance from one another in a sequence.

In transformers, words in a sentence are treated somewhat like individual pieces in a complex Lego structure. Each word, or part of a word, is like a unique piece with its own shape and context. Just as you would connect Lego bricks to form a complete object, transformers mirror this process in language, learning how individual words or word segments fit together to create a coherent sentence.

How might you visualize how the parallel attention of the transformer works? Imagine you have a huge Lego robot made from different colored bricks, each with different functions for actuation and sensing. Each brick (or part of the robot) is like a word in a sentence. Now, if you wanted to understand the whole robot, you wouldn't look at each brick one by one; you'd want to see how all the bricks connect and relate to one another at once. That's what the transformer does for sentences. Let's say you're trying to understand a specific part of the robot, maybe the robot arm. Some bricks are more crucial than others to understand that specific part. With the attention mechanism, important bricks appear bigger and clearer, while less important ones fade a bit. This helps you focus on the essential parts of the robot (or sentence). Sometimes, you want to know how one brick of the robot relates to all the other bricks. With self-attention, every Lego brick can use the magic magnifying glass to see how it fits with every other brick in the robot, similar to how each word in a sentence uses self-attention to evaluate how it fits in the written context.

The AI community has trained multiple transformer models. They can be fine-tuned and optimized for different jobs, too. The original transformer model was designed for translation tasks; basically, the researchers wanted to figure out how to improve Google Translate. Variants like BERT and GPT, on the other hand, were designed to be large language models. The BERT model learned by masking or hiding a word in a sentence, then trying to predict which word was missing. This approach helps BERT understand context and the relationships between words in a sentence. GPT is trained by trying to predict the next word or token in a sentence fragment given a sequence of words. This is closer to an advanced form of auto-complete. However, during its training, GPT learns that the most probable next word is not always the most contextually appropriate. Sometimes, choosing a word with a slightly lower probability can lead to more coherent or realistic sentences.

Imagine you have a sentence: "I love to eat ___." If you hide the word after "eat," you might guess words like "pizza," "chocolate," or "apples." You can decide whether the guess is correct or not by comparing it against the original full sentence. The same approach would apply if you included the word "apples" and removed "eat." The transformer does this at a huge scale, processing countless sentences and continuously refining its understanding of language patterns and context.

We humans learn from experience and sensory input. Transformer models learn from data. Training one involves feeding it a massive number of sentences, removing (or masking) certain words from those sentences, and asking the model to predict the missing words. Each time the model guesses, it checks against the actual word. If it's wrong, it makes slight adjustments to its parameters (the knobs) to improve its guesses in the future. The model essentially teaches itself from the data. The more diverse and vast the data, the more the model understands language's nuances, patterns, and intricacies. By looking at a lot of sentences, transformer models identify patterns that correspond to grammar, facts about the world, and even elements that correspond to reasoning abilities and common sense. Moreover, because transformers have so many parameters, they need vast amounts of data to adjust each parameter, ensuring they don't merely memorize the data. This comprehensive training process helps them generalize better to new, previously unseen data.

After training a transformer, the result is a general-purpose model. This model has seen so much text (or other types of data, in the case of transformer-based architectures for videos, etc.) that it has a broad understanding of language and how words fit together. This is the foundational model—it's like a car engine that's been run and tested, but not yet tuned for a specific race. Sometimes, we want our model to be good at a specific task, like medical diagnosis, mechanical design generation, or poetry creation. Fine-tuning is the process of training our foundational model on specific data related to that task. It's like tweaking our car engine for either a racetrack or a long-distance haul. Training the basic model is generally very time-consuming and has a high energy cost. Fine-tuning is fast.

A transformer-based tool like ChatGPT can generate coherent and contextually relevant responses, but it may also hallucinate, or generate information that seems plausible but is actually incorrect or nonsensical. To understand why, we need to consider the role of statistics in its operation. Transformers function by determining the statistical likelihood of one word following another. When generating responses, ChatGPT uses these statistical patterns to predict the most probable next word or phrase in a sequence. Its reliance on probabilities without true comprehension of content may result in a stitching together of segments from different parts of the embedding space. ChatGPT is essentially guessing based on patterns it has seen in the training data, without the ability to verify the truth or contextual appropriateness of its outputs. The model doesn't really know or understand what it's saying.

Initially, the transformer architecture revolutionized the way we approach machine learning tasks related to language, but it has now been applied to images, video, and other types of data. There is a great deal more complexity to how these models are trained and optimized. And their output is phenomenal. But I still would not rely on them to write this book! A model might be able to generate something that looks and at times reads like a book, because it has ingested and studied the patterns between words and sentences and paragraphs in many similar or even loosely related passages out there on the web. Yet it would not be representative of my ideas or opinions about the current and future states of this technology. I'm using words as a knowledge and opinion transfer mechanism between myself and you, the reader. And I need to choose those words carefully to make this work. But I have some familiarity with writing; it is an essential skill in academia. If you asked me to create a work of visual art expressing these ideas and opinions, I might defer to a different algorithm called Stable Diffusion.

STABLE DIFFUSION

Let me back up for a moment. Traditional machine learning and AI models are trained on vast datasets that can include millions of images, texts, and other data types, all labeled by humans. For instance, in image recognition tasks, intelligent humans annotate each picture, noting whether it features a beach, a car, or any other identifiable subject. The models study each digital image starting at the pixel level, gradually filtering them through neural networks of thousands of layers, until they broaden the scope of their analysis to encompass not just a few square pixels but the whole image of millions of pixels. All the while, they are picking out patterns in the arrangements of these pixels that correspond to the word "car."

Eventually, when they identify enough of these recurring patterns in images that feature cars, they develop their own conception of what that means—the many features that correspond to the presence of a car in a picture. In the vast collection of images on which we train the model, we have many instances and variations of cars, yet they all correspond to a single label. That one word, "car," represents all of those digital variations.

What generative AI does is flip this relationship. Instead of going from many images to one word, such as "car," a generative model starts with that word, and then has to decide which of the nearly infinite possible variations to generate, and in what sort of setting. This could include the type of car, its orientation or size within the image, specific colors, and much more. My MIT colleague Philip Isola likens these different variables to dials that can be adjusted to fine-tune the image characteristics. And there are hundreds or even thousands of such dials. So how do the models actually choose? We could simply let the models run and turn the dials at random, but that wouldn't be very useful as a creative tool. So instead, one of the more popular approaches relies on the Stable Diffusion algorithm.

A Stable Diffusion model is trained on labeled digital images, much like its predictive predecessors. Instead of trying to uncover the hidden patterns, however, these generative diffusion models introduce a unique process: they start digitally scribbling on the image, gradually adding noise to the image, pixel by pixel. The models randomly change the pixels and track each change as the image slowly progresses from a clear picture into a random mess of pixels. Eventually, the original image is no longer visible or distinguishable—it is destroyed. The car disappears. But the model has studied and recorded every tiny change along the route to noise. Once the model does this with enough images, it starts to notice different patterns in the noise, and correlates those subtle differences back to the original images and labels. The model learns to understand the reverse process: how to gradually remove noise to re-create the original image or generate new ones. By mastering the transition from a clear image to noise and back, the model becomes capable of generating clear images from random noise.

An intuitive way to think about the diffusion process is to imagine a confetti-packed piñata. Now picture a Major League Baseball home run champion smashing the piñata, and its tiny innards scattering through the air and all over the ground. This chaotic scene represents the starting point in Stable Diffusion's process, similar to how the model begins with noisy, unstructured data. Imagine observing the scattering process in reverse, where the confetti pieces gradually align and come together to re-form the piñata. This is akin to how Stable Diffusion learns to reconstruct images from noisy data. By studying how the piñata breaks apart into noise, the model develops an understanding of how to reverse this process. Through a series of controlled and iterative steps, what began as random noise is meticulously organized and refined, slowly emerging as a coherent and detailed image, or, in this case, the piñata made whole once again. Drawing inspiration from real images, the model carefully guides this chaotic starting point, transforming it over time. The result? After many refinements, what began as random noise emerges as a coherent image, not exactly replicating any single original, but bearing a striking resemblance to real things. This approach to data generation is deeply rooted in statistics, as the model constantly gauges how closely its creation mirrors real images and adjusts accordingly. One of the unfortunate consequences is that Stable Diffusion can create very realistic and unusual deepfakes. In 2023, the internet was abuzz with debate and discussion when a deepfake image generated by Midjourney surfaced depicting the Pope in a white puffer coat. The image stimulated important public discourse on AI-generated deepfakes, and we will talk more about the dangers of AI-generated imagery in Part Three.

Stable Diffusion can be used with different types of data. If it is trained on images, it generates images. If it is trained on mechanical designs, it generates mechanical designs. If it is trained on protein structures, it generates protein structures. When you experiment with one of these models, you can input text describing what you would like it to produce. The more specific your instructions, the more dials you're turning. The rest is randomly chosen by the AI. There are different ways to architect a model around the Stable Diffusion algorithm, so generative art programs have different strengths. The data you feed them also determines the nature of the output. They are similar to large language models in that they generate images that are the interpolation of all the data they've seen and trained on, so an artist will need to be selective and strategic if they want to use generative tools to create original work. This technology is very controversial in the creative community, too, given that many of the tools were trained in part on the creations of working artists and writers who did not grant the companies building the models the permission to do so. I hope these ethical and economic battles are resolved, because for someone like me, who lacks the skill and training of an artist, these tools are a marvel. They allow me to quickly turn an idea, captured in a few words, into an image. As technology progresses, it's exciting to ponder the new vistas Generative AI might open, both in digital artistry and in broader applications where accurate, dynamic image generation is pivotal.

10

Optimizing

IN MARCH 2016, 200 MILLION PEOPLE TUNED IN TO WATCH eighteen-time world champion Lee Sedol compete against an AI system in the game of Go. By that point, AI programs had already beaten human grandmasters in chess, but they relied heavily on brute force computation, calculating vast numbers of possible moves. The game of Go has simpler rules than chess, but its simplicity in terms of piece types and movements allows for more complex strategic decisions regarding territory control and influence. Players position white or black stones on a board, trying to surround their opponent's stones and claim territory. The number of possible configurations dwarfs that of chess, making the brute force approach impractical. To defeat a human in Go, the AI system would have to truly strategize, exhibiting a depth of thought previously believed to be uniquely human.

The company DeepMind, now a division of Alphabet (Google's parent company), designed an AI player, AlphaGo, around a deep neural network trained through reinforcement learning.

There are a number of ways AI systems learn. Some rely on human input and feedback, while others operate more autonomously. Reinforcement learning centers on training independent agents to take actions in an environment in order to maximize cumulative rewards over time. This approach trains an AI player (a computer program also called an AI agent) to continuously refine its strategies in pursuit of winning or achieving its goals.

Since these AI agents adjust their actions to increase their chances of victory, a system that adapts through reinforcement learning is going to be really good at any game that allows it to test strategies through trial and error. The agent can randomly try different paths and watch what happens. Eventually, the AI system will learn to avoid the moves that are unproductive and favor the ones that are productive. The designers of AlphaGo first had it observe games between amateur human players to give it a basic understanding of how Go is played. Then they had AlphaGo play against variations of itself thousands of times. All the while, the system was improving. While AlphaGo's primary evolution was through this self-play, there were phases of supervised learning with human input and feedback as well. Eventually, AlphaGo defeated Go world champions, including the great Lee Sedol, occasionally utilizing strategies that these masters had never imagined.

At the heart of reinforcement learning is an agent that interacts with an environment. The agent performs actions, to which the environment reacts and provides feedback through rewards determined by the effectiveness of those actions. At any given time, the environment is in a certain state—the positions of the various white and black stones on the Go board, for example. This state provides the context in which the agent decides its next action. Based on the state, the agent chooses one from a set of possible actions.

This decision-making process is driven by a policy, which is essentially a strategy. After taking an action, the agent receives a reward from the environment, which indicates how effective that action was in achieving the desired goal. This reward can be positive (if the action was beneficial) or negative (if the action was detrimental), and it is usually specified as a number, not a cookie or lollipop. The main goal of the agent is to maximize its cumulative reward over time. To achieve this, the agent continuously refines its policy based on the rewards received. This strategy may be as simple as a lookup table or as complex as a deep neural network, and the agent updates its knowledge and tweaks its approach as it plays. There is a trade-off between exploration (trying new actions) and exploitation (sticking with known, good actions). Too much exploration might lead to suboptimal short-term decisions, while too much exploitation might mean missing out on better strategies. As the agent gains more experience, it refines its policy to make better decisions and accumulate higher rewards.

We use reinforcement learning in robotics to teach machines tasks that are too complex to be programmed by humans. For example, programming an autonomous car to win a race against multiple other cars would be far too difficult. The environment is too complex and there are too many possible moves. So in one of our projects, we let the robots teach themselves through reinforcement learning instead. First, we built a simulated racecourse that incorporated the physics of the real world. If we didn't include gravity, for example, then the system might find that flying directly from the start to the finish would be faster than driving along the course. We used a dynamic, realistic model of the vehicle. With this setup, we enabled the system to explore various strategies, and tested them across thousands of simulated races.

The artificial brain of our race car was trained hierarchically, advancing from basic to complex tasks, starting with learning to steer and stay on the track and not get thrown off course. The next layer of training was about racing and going fast along the track. Another, more complex layer built in teamwork, allowing cars to learn how to cooperate in a multi-vehicle race, so one car learns to block while another blasts ahead, another learns to overtake, and yet another learns to execute a pit maneuver. Winning a simulated race yielded positive rewards, whereas losing or veering off track produced negative results. Over time, the system discovered winning techniques and learned to avoid losing strategies through trial and error.

In the real world, running thousands of races with autonomous vehicles isn't feasible. However, as long as our simulation is accurate and powerful enough, the AI brain of the cars can develop and discover its own strategies through reinforcement learning, then apply these techniques in reality. We use this approach to solve a wide range of open-ended problems. While AlphaGo stands out as a prime example, OpenAI Five demonstrated its prowess by defeating professional teams in the e-sports game Dota 2 in 2019. OpenAI took on this particular challenge because they thought that the effort would drive advancements in reinforcement learning methods. Initially, they couldn't win. Instead of reinventing the AI's learning technique, though, they simply scaled up its training. For the first competition, they'd trained the model on the equivalent of 10,000 years of human play. To get ready for the 2019 competition, OpenAI Five played against itself in parallel for the equivalent of 45,000 years. This took ten months in real time. But each day, the system was playing 250 years' worth of games, making it exponentially more experienced than the skilled human gamers it eventually defeated.

In healthcare, researchers have used reinforcement learning to design AI solutions that can analyze radiology results, identifying potential health risks that aren't detectable by doctors and technicians. My friend Manuela Veloso, who has done breakthrough work in cooperative agents through her robotic soccer competitions, has worked with a team of experts at the investment bank J.P. Morgan to develop reinforcement learning methods to optimize strategies for trading stocks. And Alphabet isn't merely using its advanced AIs to win board games. DeepMind applied deep reinforcement learning to manage power usage in Google's data centers, achieving a 40% reduction in cooling costs. As more AI tools are deployed in the real world, they will need to adapt to new data and changing environments, so I suspect that reinforcement learning will become more prevalent and better known in the years to come.

This unique approach of learning by interaction has already shown promise in many applications. Central to its success is the balance between exploration, where agents test novel strategies, and exploitation, where they stick to known successful methods. While powerful, reinforcement learning is not without challenges. It primarily depends on vast amounts of experience rather than the large datasets from which predictive AI derives its power. In supervised learning, models are trained on labeled datasets, where more data often leads to better performance. But reinforcement learning agents learn from interacting with their environment, making decisions, and receiving feedback in the form of rewards. This means that these agents often need to explore a large number of states and actions in the environment to learn a good policy, which can mean that they need substantial computational resources. If the environment changes in a way the agent hasn't experienced before, the agent can struggle because it hasn't learned the right actions for this new situation. For reinforcement learning applications in the physical world such as robotics, the move from virtual to physical, or what we call the sim-to-real transfer, is crucial. Safety in reinforcement learning is important, especially in real-world scenarios where incorrect actions could have dire consequences. Furthermore, when multiple agents interact within an environment, as seen in e-sports and in our car racing project, the complexity rises, requiring agents to predict not just environmental responses but also the actions of their peers. While there are limitations, reinforcement learning has cemented its role as a pillar in the world of AI.

11

Deciding

AI ENCOMPASSES A VARIETY OF ALGORITHMS AND TECHniques beyond machine learning. Among these alternatives, decision-making AI methods are critical. The distinction between decision-making and machine learning is fundamental; they serve different purposes and operate based on different principles. At its core, machine learning is about training models on data to make predictions or classifications. Machine learning depends on data and extrapolates predictions without explicit programming, answering questions such as, "Based on what you've learned from the data, what do you think will happen next?" Or, "Which category does this new piece of data belong to?" Decision-making evaluates multiple potential solutions to a problem and selects the best available option based on different criteria or metrics. It answers questions such as, "Given these options and outcomes, which choice is the best?" For instance, in a game, the goal is to win, and the alternative that offers the best possible chance of reaching that goal will determine the final decision. There's still some interconnection between them; many decision-making models leverage machine learning components to help evaluate or rank those potential solutions.

One of the key distinctions with decision-making AI systems is that they can make their suggestions based on predefined rules or logic. This rule-based approach prevalent in early AI research was termed GOFAI (good old-fashioned AI). And in certain instances, it works quite well! In the classic 1983 film *WarGames*, the heroes prevent an apparently self-aware AI from destroying the world by coaxing it into playing a never-ending sequence of tic-tac-toe games against itself. In reality, with rule-based logic AI can easily play this game flawlessly. And if both contestants were to play flawlessly, then the outcome would inevitably be a draw.

So let's imagine you were to play tic-tac-toe with the help of a decision-making AI. Your opponent enters an X in the upper left square. Now let's say your AI system is built around the Minimax algorithm, a classic in decision-making systems often used in gameplay. The algorithm's goal is to pick your best move by exploring all your possible moves and the subsequent counter-moves by your opponent, potentially analyzing every possible game outcome. For each possible move, the AI system assumes your opponent will respond with the move that's worst for you (hence "minimizing" your outcome), and that you will always want to choose the move that's best for you ("maximizing" your outcome).

This exploration of possible futures is called a look-ahead, and at the end of the look-ahead—which can either stop at a predefined depth (in a tic-tac-toe game that has already started, the remaining depth would be eight moves, since there are only that many squares left) or extend all the way to the end of the game—the system evaluates the board's state.* For terminal states like a win, loss, or draw, it assigns a score. This score is then propagated backward through the previously explored game states, helping to determine the value of each move. Finally, the AI system suggests a move that will most likely result in the best outcome. Basically, the decision-making AI will suggest the move that gives you the highest probability of victory at each point in the game.

The major difference here, relative to the kinds of AI solutions we have been discussing throughout this book, is that the system wouldn't be learning or improving its performance as it assists you. A system like this would be programmed with the rules of the game and the algorithm to compute optimal moves. Then it would map those strategies to what is happening on the board. The system doesn't need a learning process in order to make informed decisions.

This doesn't merely apply to decision-making AIs based on the Minimax algorithm. The same holds true for other influential solutions in this space. Dijkstra's algorithm, for instance, finds the shortest path in a network, and has been used for everything from identifying potential friends in a social media app to identifying the ideal path between two points on a street map. The branch and bound method focuses on optimization problems, seeking the best solution by systematically pruning paths that cannot possibly yield a better solution than the best one found so far. Then there's the Hungarian method, which businesses rely on to solve assignment problems, for example to figure

* The basic Minimax algorithm can be computationally intensive as it explores all possible moves, but it's often enhanced with pruning tools that allow it to skip evaluating branches that clearly won't lead to victory.

out how to allocate various tasks to workers at different pay scales in a way that optimizes costs and efficiency. Imagine you have three employees, and three tasks that need to be completed, but each of your workers commands a different wage for the varied assignments. You could sit down and calculate each of the possibilities manually, but this would be arduous and time-consuming at an organizational scale. The Hungarian algorithm rapidly calculates a solution, suggesting how to assign the work in a way that minimizes cost.

Generally, these algorithms and others like them offer systematic or heuristic approaches (i.e., strategies that provide a quick approximation solution) to decision-making, enabling users to make near-optimal choices in complex scenarios. They aren't necessarily ideal for making predictions or improving their performance as they gain experience, but they can be powerful allies in many decision-centric scenarios, including medical diagnostic applications, business strategies, or gameplay. There are also instances in which we enlist the help of machine learning in decision-making processes, especially in complex scenarios that call for something more than traditional rule-based approaches.

Returning to that game of tic-tac-toe, machine learning algorithms would attempt to learn patterns or strategies from past game data and use them to make decisions in new games. They'd adapt based on the data. Or if we were to consider a different game, such as chess, a machine learning solution would start by looking at a series of recorded games, analyze sequences of moves from this corpus, evaluating board positions and inferring effective strategies from this data. The system would attempt to back out strategies based on existing data. Then it would implement these strategies in new games and adapt based on how it performs. An algorithm like Minimax doesn't learn in that traditional sense; it doesn't adjust its strategy as a result of past experience. Instead it looks at all possible outcomes, making decisions based on logical reasoning and the game's inherent rules. While Minimax is foundational for chess due to the game's structured nature, in practice it's often combined with other techniques to achieve elite-level performance.

The Minimax algorithm, as used in games like chess, offers a systematic way to evaluate potential moves by looking ahead and considering all possible outcomes. This logical, branching methodology of weighing various decisions is related to a foundational decision-making tool: the decision tree. Much like the tree of possibilities Minimax generates for game moves, a decision tree visualizes choices and their consequences. Starting from a root decision or observation, the tree branches out, delineating possible outcomes and subsequent decisions or predictions. When presented with a new data point, the system traverses down the tree, following the branches corresponding to the data point's attributes, and arrives at a leaf node, which provides the decision. While Minimax is optimized for competitive games with a clear adversary, decision trees generalize this approach, making them suitable for a wide range of scenarios. They not only help in making a decision but also allow us to understand the reasoning behind the decision, providing a transparent and structured way to untangle complex decision spaces.

With larger questions involving massive datasets, we might turn to a more powerful variation, the Random Forest algorithm. Instead of building one decision tree, the Random Forest approach selects subsets of the data and applies a decision tree to each of them, then takes the average result of those decision trees. The selection process is random and there are a great many trees. Hence, random forest.

Because they are based on binary choices, decision trees fall short in scenarios rife with uncertainty and complexity. In such situations, we turn to a more nuanced tool: Bayesian networks. These networks capture individual decisions or events and also the probabilistic relationships between them. They are rooted in Bayes' theorem and allow us to account for prior knowledge and adjust our understanding as new evidence surfaces. Imagine a weather forecast saying there's a 50% chance of rain. You don't feel like carrying an umbrella, so you take your chances. Before you leave the house in the morning, though, you see dark clouds. Now you're more certain of rain. Bayes' theorem is about adjusting probabilities based on new evidence—in this case, updating the likelihood of rain due to the sight of dark clouds. Bayesian networks are used in complex domains where variables have intricate interdependencies, such as medical diagnosis or risk assessment. By capturing the probabilistic relationships, they provide a structured and quantifiable way to reason under uncertainty, making them a powerful tool in the decision-making toolbox. A tree gives us a clear pathway from root to leaf, while a Bayesian network paints a broader picture, weaving a web of interrelated decisions and outcomes, each influencing the others in a dance of probability and inference.

• • •

When the team behind IBM's famed Deep Blue was training their chess system, they tested it against grandmasters, and in the early years, the machine often fell short. Although Deep Blue didn't learn from those early mistakes in the way contemporary machine learning systems do, the human researchers did. They studied the games, identified the system's shortcomings, and then reprogrammed it to rectify the errors. Additionally, they studied various other games and grandmaster strategies and programmed these into the system. When Deep Blue beat chess legend Garry Kasparov in 1997, the system was able to search anywhere between 100 million and 200 million alternatives every second as it considered its next move, or decision. The system wasn't smarter than Kasparov, but its computational prowess allowed it to consider a vast range of potential moves and map these to known chess-playing strategies.

After that landmark victory, subsequent efforts to develop AI systems capable of winning against people shifted toward machine learning. IBM's Watson, which triumphed at the game *Jeopardy!*, utilized a blend of natural language processing and knowledge representation, while Google's AlphaGo, which mastered the ancient game of Go, heavily employed deep learning alongside a tree search technique. AI solutions focused on decision-making remain indispensable today. However, it might be more appropriate to regard these systems as offering suggestions rather than actually making decisions. From the structured logic of Minimax in games, to decision trees and the probabilistic insights of Bayesian networks, these tools assist people in navigating intricate situations. They aim to mirror the human process of evaluating options, gauging consequences, and adjusting to fresh insights. As technology progresses, games like *Jeopardy!* and Go become more than entertainment; they serve as the benchmarks of AI's computational prowess. But true, impactful decision-making has to be intertwined with human intuition, experience, and values. These AI tools are best seen as assistants and cognitive aids, informing our decisions rather than dictating them. The final judgment should lie with people, not algorithms.

A Business Interlude

The AI Implementation Playbook

THE AI TRANSFORMATION UNDERWAY TODAY IS AS important as the digital transformation of two decades ago. AI-native companies will have an advantage similar to those enjoyed by digital-native organizations—they will be able to embrace and deploy new technologies faster and enjoy the rewards sooner. Everyone else? Your workforce, infrastructure, and strategy will need to be adapted. Yet I would not advocate wholesale adoption of every buzzy new AI solution that appears on the market. There is risk in moving too slowly and falling behind your competition, but moving too quickly could be equally problematic.

The key to getting this transition right is mapping out the specific needs of your business, designing and implementing an AI transformation strategy that directly addresses your pain points while maximizing value, and acquiring the talent necessary to make this transformation work for your organization now and into the future. In this new world your personnel might be more important than ever. As you evaluate how AI can help your business, I'd suggest breaking down the various roles within your organization by task, then looking at whether AI can assist, augment, or automate those tasks.

ASSIST

In this mode, AI systems are deployed to support human operators, helping them make better decisions, often in real time, and to complete tasks faster with more ease. For example, while humans can analyze data, AI can process vast amounts of data quickly, revealing patterns and insights that might go unnoticed by human analysts, helping with decision-making processes in areas like marketing strategies, financial forecasting, and operations.

AUGMENT

Here AI is not so much an assistant as a tool that enhances human capabilities beyond our natural limits, making us more efficient and effective. The programming assistant Copilot is one great example. Another is the way in which creative advertising agencies are increasingly using deepfake technology for clever branding campaigns. In India, Pepsi created a popular video featuring two versions of the famed actor Salman Khan, and the gambling company FanDuel used generative AI to create a younger version of the former basketball star and current commentator Charles Barkley. The deepfake and the real Barkley bantered casually on the couch as they promoted the brand. Both ads show how teams can use AI to augment their creative capabilities and make something new and memorable. Yet for me this approach is best summed up with a medical research example, which demonstrated that while AI might be very good at a specific diagnostic task, and human experts slightly better, the most impressive results are achieved when people and AI work together. In this experiment (also described in my book *The Heart and the Chip*), medical doctors and an AI system were asked to review images of lymph node scans and classify them as cancerous or not. The human doctors achieved the task with 3.5% error, while the AI system returned a 7.5% error rate. But working together, the doctors and the AI system achieved 80% accuracy improvement to 0.5% error, because humans and machines see different things in different ways.

AUTOMATE

Automation is when AI systems can take over tasks entirely, replacing human involvement. This can range from simple scheduling to autonomous driving. For functions like finance or operations, AI can generate routine reports without human input, ensuring stakeholders receive timely updates. In the insurance industry, startups are developing tools that automate data entry, freeing adjusters to work on higher-value tasks.

• • •

As you look across your organization, you will need to determine which tasks could fit in which categories. This is not an easy job, and to do it right you will need to hire a new class of professionals: bilinguals. I'm not talking about individuals who are fluent in multiple human languages. You need people with a deep understanding of AI *and* your business. For example, my husband's company, Codametrix, is automating medical coding, which is necessary for insurance reimbursement, but happens to be a laborious and time-intensive process that often falls to doctors and other overqualified healthcare professionals.

This code identification and entry task is an ideal target for automation. First, because it is possible. Second, because it frees up the far more valuable time of doctors. Yet that doesn't make operating this particular business easy. The world of medical coding is undeniably niche and the technology required to automate the process undeniably complex. So Codametrix needs people who understand *both* the medical coding world and the AI world to test and evaluate performance. The machines do not deliver 100% accuracy, so these bilinguals also help determine when a human needs to take over and do the coding because the machine has too much uncertainty in its computation.

Regardless of your industry, you will need these people, and they are going to be extremely valuable in the economy at large. As a business leader, you will also need to educate yourself on these AI solutions, foster a culture of continuous learning and adaptation to nurture more bilinguals within your organization, find collaborators and partnerships who broaden both your toolset and knowledge base, and perhaps work with universities to understand what technologies are coming around the corner. What you need to know will be shaped by your role within your organization, and whether you are charged with leading, using, deploying, or developing AI solutions.

Adopting AI is no minor shift. Transitioning from theoretical benefits to practical implementation requires systematic planning, appropriate resources, and a clear understanding of the goals and limitations of AI. Recognizing this complexity, several colleagues and I developed a comprehensive template of questions and steps to guide the design and implementation of AI solutions in large organizations, as part of a sequence of courses on AI for national security leaders and AI for business. Retired General Stephen "Seve" Wilson of the U.S. Air Force and Diane Staheli of the MIT Lincoln Laboratory came up with the original idea to create what they called "blueprints for action." Anu Myne and Robert Bond of the MIT Lincoln Laboratory took the concepts and crafted the actual blueprint material, and then we worked as a group to build the class. The blueprints for action have evolved alongside the rapid advances in AI and our cohorts of learners. The following section, adapted from that template and the resulting classes, is meant to give you a sense of the nuances of AI implementation. The goal is to help ensure that your approach is not only technically sound but also strategically aligned with your business objectives.

The first question is whether AI is right for your business in the first place.

ECONOMICS & STRATEGY

Before you go deep into an AI project, you will want to forecast the costs and benefits and how they will evolve over the next few years. Scale is critical. If you only have a small number of people performing the task you're considering for automation, then it will probably not be worth the effort to deploy an AI solution. Yet if the task is necessary across multiple organizations or industries, then you can think about adopting a platform strategy and selling the service to others.

Let's say you clear this first hurdle. Now what?

1. DEFINE CLEAR OBJECTIVES

The next step is to figure out what you want AI to *do* for your business. This could range from automating routine tasks, like processing customer service inquiries through chatbots, to enhancing accuracy in quality control through image recognition systems, or even discovering new market insights by analyzing consumer behavior data. Here you should determine whether you really need AI. You must be certain that AI adoption is a true strategic enhancement tailored to your unique challenges.

BUILD OR BUY. At this point, you will need to address whether to purchase the needed solution from another vendor or dedicate the resources to build it yourself. This 'build or buy' decision should be informed by a thorough assessment of your specific AI needs, market research, cost-benefit analysis, and consideration of the potential broader applications and benefits of a custom-built solution. As you research off-the-shelf AI products or platforms to see if they align with your specific requirements, consider factors like compatibility with your current systems, scalability, and customization potential. If a market solution meets your needs with minimal adjustments, buying might be the more efficient and cost-effective route. However, if the market doesn't offer a satisfactory solution, or if the available products require significant modification to suit your needs, then building a custom AI solution may be preferrable. This path allows for tailored functionality, complete control over the development process, and the opportunity to create a solution that perfectly aligns with your business processes and objectives. However, it also involves considerable investment in terms of time, resources, and expertise. An interesting aspect to consider is the potential of converting your custom AI solution into a new business venture that could be commercialized and sold to other companies. For instance, an AI-based inventory management system developed for your company could be adapted and sold to other businesses that rely heavily on stock control, like retail chains, manufacturers, or logistics companies. This not only creates a new revenue stream but also positions your company as an innovator in the AI space.

DETERMINE EXPECTED RETURN ON INVESTMENT (ROI). Next you will need to create an expectations roadmap with specific measurable goals, including the potential return on investment (ROI). An implementation that leads to marginal gains may not be worth the investment. Here it might prove helpful to keep in mind that not all tasks which can be automated should be automated. In some cases, the cost of developing or adopting an AI solution to fulfill the task may be too high. Ideally you want a clear sense of what you expect from this investment in AI before you get too deep into the project.

DETERMINE THE STAKEHOLDERS. Next you should pinpoint the individuals, including bilinguals, who will use and manage the AI deployment, and the customers (both external and internal groups or departments within your company) who stand to benefit. These could be employees who will interact with the AI tool daily, such as customer service representatives using an AI chatbot, data analysts working with AI-driven analytics, or factory workers managing AI-automated machinery. Understanding their workflow, challenges, and expectations is key to designing an AI system that enhances their productivity and job satisfaction. It is also important to consider the managers or overseers of the AI deployment, because these individuals are responsible for monitoring and maintaining the AI system. Their insights will be critical for defining system requirements, integration with existing workflows, and ongoing management needs.

DEFINE THE VALUE PROPOSITION. You should be very specific about the proposed enhancements as well—what the potential AI solution will offer that other solutions or people might not. Evaluating the potential customer experience is critical as well. (I dislike incompetent chatbots and robotic call services.) Once you identify areas of potential improvement, focus on how to maximize these benefits by choosing the appropriate AI techniques and determining whether they will be used to assist, augment, or automate.

DETERMINE PARTNERS AND ROLES. The next step is to identify the partners involved in your AI project. This includes partners for (a) adapting commercially available AI applications to your specific needs, (b) modifying existing AI techniques for your use, and (c) creating entirely new AI solutions tailored to your requirements. Additionally, you must determine who will collaborate with you in forming an effective human-machine team. It's crucial to specify the roles for evaluating the scope, safety, and ethical considerations of your AI initiatives, and define who will be responsible for collecting, evaluating, organizing, storing, and overseeing the data that will feed your AI systems. You'll need to establish clear guidelines on data sharing, including who will have access to this information, and assign a team or individual to manage these aspects, addressing any questions and concerns as they arise.

Implementing AI is more than just introducing new software; it's a powerful technology that interacts with your systems and impacts your people. In addition to cultivating a team of business bilinguals, you might consider partnering with specialists who can aid in designing, building, and maintaining a workforce that is equipped to work with AI technology, emphasizing the need for a diverse team with varied skills and roles focused on safe, efficient, and effective AI utilization that benefits both your organization and your people.

DEFINE MEASURES OF SUCCESS. How will you know if it works? In our class, we recommend establishing a comprehensive set of performance measures. You might start by identifying key metrics to assess the envisioned AI capability, such as accuracy, efficiency, and impact on business outcomes. In the context of human-machine teams, you should evaluate how the AI applications enhance team productivity, decision-making, and overall job satisfaction. You'll need to ensure that AI decisions are transparent, fair, and in compliance with regulatory standards. Data integrity and quality are fundamental to AI effectiveness, so you'll need robust techniques for data validation and ongoing quality checks. Additionally, you should rigorously assess the SAFER (safety, assurance, fairness, explainability, and robustness) attributes of your AI systems. This can involve regular audits, bias detection mechanisms, and testing for resilience under various scenarios.

DEFINE BUY-IN, SUPPORT. While some AI tools might integrate smoothly into existing processes, a comprehensive AI transformation often involves tackling more complex and less-obvious use cases. Gaining buy-in and support for an AI application within your organization is a critical step that involves more than just showcasing the technology's capabilities. You need to be able to demonstrate how the AI application aligns with the company's goals, addresses specific challenges or opportunities, and how it will impact your people. The complexity of the implementation process can be surprising. For instance, you might need to deploy sensors or other equipment—a scenario I review in detail in chapter 15.

You will need to evaluate whether you have the computation infrastructure to support a given AI model, and whether the application can be supported through on-site hardware or if it needs to enlist a cloud service. Finally, there will be a significant human component to the effort. Your people are going to want to know how this technology will affect them. You will need to achieve buy-in with the user community as you introduce this new AI-based capability. That might involve identifying allies and advocates and strong communications efforts explaining why the solution is being deployed and how it will help both the organization as a whole and the people tasked with working with these tools.

DEFINE THE RISKS. During this process you will need to define the risks and limitations of employing an AI solution. Data quality and representativeness must be thoroughly evaluated. A team of experts should scrutinize the training data for issues such as bias, overfitting, or lack of generalizability. This is crucial because the quality of the data directly influences the performance and fairness of the AI model. If the training data is not representative of the diverse scenarios the AI will encounter in real-world applications, it may lead to inaccurate or biased outcomes. You will also need to explore the potential for unintended consequences. For example, AI chatbots designed for customer service might struggle with complex or nuanced queries, leading to increased customer frustration. It's essential to have strategies in place to identify and rectify such issues promptly, ensuring a balance between automated and human-driven processes where necessary. In the same vein, you will need to address data protection and privacy, establish support for multilevel security, and gain a deep understanding of the data required for your initiative and whether you have the proper permissions in place to access and compute on this data, using either your own infrastructure or a cloud service. The environmental impact of the compute operations should be evaluated, too—these solutions can be energy hungry. During both the training phase and operational use of the AI system, robust cybersecurity measures must be in place to protect against potential breaches that could compromise performance or sensitive data.

BUILD A BILINGUAL TEAM. I cannot stress enough the importance of bilinguals. Building and cultivating this AI-ready workforce is critical.

MAP OUT THE DEPLOYMENT. You will need to outline a strategic approach to deployment that considers people, ethics, and the technical requirements necessary for successful implementation and operation. Integrating the human element into AI deployment is essential for ensuring smooth adoption and effective collaboration between AI systems and human workers. This involves designing user-friendly interfaces, providing comprehensive training to employees, and establishing clear protocols for human-AI interaction. On the ethics front, you will need to ensure fairness and lack of bias in AI decision-making, transparency in how AI systems reach conclusions, and respect for privacy and data protection laws. Determining the compute requirements includes assessing the necessary processing power, storage capacity, and network capabilities needed to run the AI applications efficiently.

DEFINE A DATA PLAN. You should create a holistic strategy that details the types of data needed, how they will be collected—be it continuously, periodically, or as a one-off event—and how they will be processed and used to train the AI model. Storage and management of this data must be secure and compliant with relevant regulations, while also being adaptable to the changing needs of the AI system. Anticipating and mitigating adversarial threats, such as data tampering, will be crucial. Involving end-users in the development process will ensure the AI solution meets practical needs, and considering whether the data requires labeling or if simulated data can be employed is vital for effective training. Incorporating red team* activity will ensure robustness. Integrating AI DevSecOps† with traditional processes will lead to a seamless and secure development lifecycle.

UNDERSTAND THE RESOURCE REQUIREMENTS. Infrastructure and resources required will vary significantly based on the type of AI capability being developed. Enterprise AI might require robust data processing and storage solutions to handle large volumes of business data, whereas operational AI could need more real-time processing capabilities and edge computing resources. Project dependencies are crucial in shaping the timeline. For instance, if the AI capability relies on integrating with existing enterprise systems, the readiness and compatibility of these systems become critical dependencies. Similarly, dependencies on external vendors for specialized AI components or data sources can influence the timeline.

Dedicating resources to training and supporting your people will be vital for enhancing the effectiveness of the human-machine team. AI capabilities involving direct human-machine interaction may require specialized resources like biometric sen-

---

* A red team consists of experts tasked with breaking the system, in order to identify problems with the system.

† DevSecOps (Development, Security, and Operations): DevSecOps is a software development approach that integrates security into the entire development and operations (DevOps) lifecycle, emphasizing the importance of security at every stage of the process.

sors or customized interfaces, such as voice recognition systems for natural communication and wearable devices that monitor human physiological responses for more adaptive AI behavior. The timeline for data acquisition is another critical factor. The time required depends on the data's availability and the processes involved in collecting, cleaning, and organizing it for AI training. Lastly, you will need to determine the compute capability required for developing AI, which depends on the complexity of the AI models and the scale of data processing, and the projected costs of training models. This is high right now and in some cases prohibitively expensive. Yet it will likely come down over time.

Finally, you should detail the environments in which the hoped-for capabilities will be tested, evaluated, validated, and verified. And I cannot stress enough the importance of having internal experts and teams or external partners to whom you can address such questions.

2. COLLECT AND PREPARE DATA

Once you move through this admittedly rigorous first phase, you will start to look at how to put this plan into action. Pulling together relevant data that meets all regulatory, legal, and compliance requirements comes first. There are many technical solutions to help with this step, but you will need the right people to oversee it, too.

3. CHOOSE THE RIGHT AI MODEL

This decision will not be resolved through a simple Internet search. You will need knowledgeable people to help you select an optimal model for your requirements—one that is in line with your cost and sustainability requirements.

4. EXECUTE DEVELOPMENT AND TRAINING

Once you have your larger pool of data, you will want to divide your data into three different sets for training, validation, and testing, and then initiate the training.

5. EVALUATE THE MODEL

After training and validation are complete, you'll want to find out whether the model actually works. This is the reason for the test dataset. There are several relevant metrics here, including accuracy, F1 score,* and root mean square error (RMSE),† which helps you evaluate the accuracy of a machine learning model's predictions.

6. DEPLOY

If the model passes the evaluation test, then you will start to work it into your previously defined process, whether you want it to assist with, augment, or automate a task.

7. MONITOR AND MAINTAIN

This is not a set-it-and-forget-it operation. These are intelligent systems capable of learning and improving, and of making mistakes, too, so continuous monitoring is crucial. Plus, your busi-

---

* The F1 score in machine learning is a metric used to evaluate the accuracy of a classification model. It is calculated as the harmonic mean of precision and recall. It is useful in situations where there is an uneven class distribution for the classification problem.

† The root mean square error (RMSE) is a standard way to measure the error of predicting quantitative data. It is computed as the square root of the average squared differences between the predicted and actual values.

ness will change. Market trends will shift. So you will need to monitor, update, and potentially retrain the model over time.

8. ITERATE AND IMPROVE

You should establish policies and best practices to regularly reassess the model's performance and the objectives it is designed to achieve. You want your solution to improve over time as it interacts with real-world data.

9. MEASURE BIAS AND FAIRNESS

While you will account for bias and fairness as you compile the data in the earliest stages of the process, this should be an ongoing focus area. Any AI solution you deploy must be ethical and unbiased. Ideally, you want transparency with regards to the model's decision-making processes as well.

10. COMMUNICATE AND EDUCATE

All of your stakeholders, from end users to management, should understand how and why AI is being incorporated into your business. Anxiety over AI is very real. Misinformation is rampant. The more you can do to help people within your organization understand your chosen tool and how they are designed to help both the organization and its people, the better.

11. MAINTAIN A FEEDBACK LOOP

Your people will know how your AI solution is working in practice, so you will need to establish mechanisms and communication channels for users or systems to provide feedback on AI predictions or decisions. This feedback will help you refine and improve the AI model.

12. PREPARE FOR THE LONG TERM

This is not a quick or easy process. Yet following these guidelines will increase your chances of a successful and efficient implementation. At the same time, each application or domain might have its own nuances, so flexibility and adaptation to specific situations are key.

• • •

Let's look at an example; imagine you are operating a hospital. One of your persistent problems is readmissions. When a patient is released from the hospital and returns within thirty days, this is not only costly but hints at quality-of-care issues, so you want to use AI to predict the likelihood of patients being readmitted within that timeframe to improve your care.

1. DEFINE CLEAR OBJECTIVES: You want to find those high-risk patients and intervene before they're readmitted.

2. COLLECT AND PREPARE DATA: Start with electronic health records (EHR) as your dataset. Given the sensitive nature of this data, implement stringent security measures with strong encryption, plus regular audits. Data should also be cleansed and evaluated for bias and rebalanced to ensure fairness both in the dataset and the model itself. For example, bias in hospital readmissions could occur if the predictive model is trained on a large number of low-income patients who may not have adequate access to medication and other healthcare resources. If these conditions lead to higher readmission rates in the training data, the model might overestimate the risk of readmission for low-income patients because it learned from a dataset wherein this group had disproportionately high readmission rates.

3. CHOOSE THE RIGHT AI MODEL: Different types of data and tasks require different types of models, each with its own strengths and weaknesses. EHR data may harbor intricate relationships—for instance, how the interplay of various symptoms, history, and medications might result in a certain outcome. Simple linear models might not capture the deeper, non-linear patterns in the data. Thus, more advanced models like gradient-boosting machines or deep neural networks should be considered. Gradient-boosting machines are a type of machine learning model that builds an ensemble of decision trees in a sequential manner, so that each new tree tries to correct the mistakes of the previous ones. Alternatively, deep neural networks may be able to discern non-linear relationships and patterns in the data that simpler models might miss.

4. EXECUTE DEVELOPMENT AND TRAINING: Next, you should divide the data into training, validation, and testing sets, and then begin training on the first dataset. Typical splits are 70, 15, 15 percent, respectively.

5. EVALUATE THE MODEL: Here you assess the performance using the appropriate metrics. In this case, Area Under the Curve (AUC)* would be a good way to measure the model's effectiveness.

* Area Under the Curve (AUC) is a measurement of the performance of a classification model. It is used in machine learning to assess the accuracy of models in predicting binary outcomes.

6. DEPLOYMENT: Now put the model to work in the hospital's IT system. Here is where your bilinguals will play an important role, as they will be able to combine their technical knowledge of the model with an understanding of patient cases and the hospital as a whole.

7. MONITOR AND MAINTAIN: As the model operates, measure the validity of its predictions and retrain it as needed.

8. ITERATE AND IMPROVE: Let's say someone discovers that a specific medication is associated with higher readmission risk. Update the model so it will be more likely to identify patients on those medications as likely readmission candidates.

9. MEASURE BIAS AND FAIRNESS: The model must not discriminate against any patient demographic, so these checks will be regular and ongoing.

10. COMMUNICATE AND EDUCATE: Medical staff are trained on the new AI tool to better understand its predictions and the relevant interventions. These internal bilinguals help keep everyone informed, including hospital leadership.

11. MAINTAIN A FEEDBACK LOOP: The human caregivers track and report on the model's effectiveness and impact.

12. PREPARE FOR THE LONG TERM: The impact of a tool like this one might not be immediate, as it will need to be optimized and retrained over time, and there will be a learning curve for hospital staff and medical professionals. But if there is buy-in across the organization, and these professionals commit to providing feedback that helps to train and improve the model, then we can expect that your hospital will see a significant decrease in thirty-day readmissions within a year. This would not only demonstrate the effective practical application of AI, but reflect positively on everyone involved, from the manager or executive leaders who approved the project to the medical professionals who used the tool to improve patient care.

PART THREE
Stewardship

WITH ITS POTENTIAL TO TRANSFORM INDUSTRIES, redefine efficiency, and revolutionize the way we live and work, AI promises extraordinary benefits. However, this promise also comes with numerous challenges and responsibilities. The rapid integration of AI into science, business, education, and everyday life highlights the need for a new type of stewardship to ensure that these technologies are developed and deployed in a manner that is ethical, responsible, and aligned with human values and societal norms. Effective AI stewardship will maximize the positive impact of these tools while minimizing potential harms. It encompasses a broad range of considerations, from privacy and security to fairness and transparency, as well as the long-term implications of AI on employment, social structures, and human behavior.

As AI systems become more complex and autonomous, it is important to establish robust frameworks for AI accountability and governance. This entails not only setting standards for the ethical design and deployment of AI but also creating mechanisms for monitoring, certification, and regulation that can adapt to this rapidly evolving landscape. The goal is to foster an environment where innovation thrives, but not at the expense of ethical considerations or societal well-being. Embracing AI stewardship is crucial for steering the development of artificial intelligence in a direction that benefits all of humanity and our planet. The journey toward responsible AI is a collective one, requiring the engagement of policymakers, technologists, businesses, and the public to ensure that AI serves as a force for good.

12

The Dark Side of Superpowers

AN UNUSUAL VIDEO OF UKRAINIAN PRESIDENT VOLODYmyr Zelenskyy surfaced not long after the Russian invasion of his country in March 2022. In the clip, the president appears to encourage his people to surrender. Yet the figure in the video was not Zelenskyy, but the digital output of an advanced AI model. The clip was a deepfake, a synthesized video produced via deep learning models. This was not entirely a surprise to the president and his team; they expected that state-sponsored deepfakes might be used as a form of information warfare. They had already warned the public, and the president publicly called out the fake and reaffirmed his nation's commitment to defending its land and people.

That clip was not very convincing, but deepfakes are only going to get better. My friend Hany Farid, a computer scientist and digital forensics expert at the University of California, Berkeley, has noted that this technology has advanced from low-resolution variations to incredibly sophisticated deepfakes that are not only remarkably convincing but also present significant challenges in detection, often requiring advanced analysis techniques or expert scrutiny to reliably distinguish them from genuine content.

This book has been largely optimistic; now we arrive at the scary part. I've been focused on how AI can benefit all of us, but individuals, organizations, and states with malicious plans will also access these incredible tools. Deepfakes will continue to get even better, potentially acquiring new capabilities such as real-time generation and adapting to interactive scenarios, blurring even more the line between reality and fiction. The technologies that endow us with mental superpowers like speed, knowledge, insight, creativity, and foresight can also be misused, and it would be irresponsible to celebrate the many potential benefits of AI without balancing that vision with a clear discussion of the potential risks. This conversation and the ensuing action are essential; we need to discuss what could go wrong, and when, and start planning and implementing various steps to mitigate both known and unknown risks. The ideas below are already out there in the public sphere—I'm not inventing these scenarios or giving malicious actors potential ideas. In a sense, these are the obvious cases.

Let's start with speed. The same approach used to accelerate the discovery of therapeutic drugs and treatments could help bioterrorists develop new forms of chemical warfare. Researchers have already demonstrated that one AI system can produce 40,000 possible toxic agents in just six hours. The same increased computational speed and automatic coding that's helping programmers and introductory computer science students will aid hackers. In the movies, we typically see the hooded hacker trying one password at a time and thinking deeply about each attempt. But in reality this approach doesn't work, as passwords are too complex; with AI, hackers can get more creative. To demonstrate the potential for harm, the security firm Check Point Research used a large language model to write a phishing email—the sort that convinces recipients to click on a link or download a file containing malicious code such as ransomware. In this case, hackers wouldn't need to learn the intricacies of coding and malware development. They could capitalize on the ease-of-use of the AI to launch attacks with unprecedented speed.

The ability of these systems to swallow up so much of the world's knowledge can be a fantastic tool or a severe danger. AI can be used to scrape personal data from online sources. AI-based facial recognition technology has been increasingly reported as a tool in mass surveillance systems around the world, raising concerns about privacy and individual rights. And then there's the risk of flooding the world with false knowledge. Remember: ChatGPT hallucinates. The model generates text that sounds right, but it doesn't know the difference between fact and fiction, so using such tools to produce nonfiction articles or books is inherently flawed. Yet websites and blogs packed with ChatGPT content have already emerged, and self-published travel guidebooks brimming with false information and buoyed by fake reviews have been popping up on Amazon. Most people still trust the veracity of content within physical and digital books, and malicious actors could leverage AI to rapidly generate works that espouse conspiracy theories, and give hateful or racist ideas the appearance of legitimacy.

The AI tools we're using to help us uncover insights that may lead to new scientific discoveries or business breakthroughs can also be used to help malicious actors pinpoint people who might be vulnerable to a scam, conspiracy theory, or covert political campaign. In 2013, researchers showed that digital records of a person's Facebook likes could be used to predict their sexual orientation, ethnicity, personality traits, intelligence, and many other sensitive personal attributes. Later, in 2018, journalists revealed that the firm Cambridge Analytica had derived insights from data that allowed it to identify and target potential supporters of presidential candidate Donald Trump, influencing the outcome of the election. And this was before the advent of generative AI. Which brings us back to the risk that garners some of the most attention—the threat of deepfakes. The same tools that can produce photorealistic images of dogs playing guitars on the surface of the moon, or humorous videos of an AI-generated version of the actor Tom Cruise, can be exploited for truly frightening ends. The Zelenskyy example is one of many. We are developing various technological solutions to detect deepfakes, but these may only lessen, not eliminate the damage.

Our ability to use AI and machine learning to enhance our foresight is another double-edged sword. Bad actors could use these tools to predict troubling economic trends and turn disasters into financial windfalls at the expense of others. In 2010, a trader deployed a particular set of algorithms that allowed him to rapidly place and then cancel thousands of orders, which led to one major financial market dropping by nearly 10% during the day, only to recover much of its value before trading closed. Or imagine a group intent on influencing an upcoming election developing an AI system that predicts how the public might react to different breaking news stories or possible scandals, then, with the help of generative AI, manufacturing an incident tuned to their findings in order to shift public opinion toward or against a particular candidate. The use of AI to anticipate and manipulate public reactions, and launch misinformation campaigns tailored to be maximally effective based on real data and predictive analytics, all of it performed on a large scale—this kind of work poses significant threats to democracy itself.

If AI helps all of us master new skills faster, that includes the less principled among us, too. The security leader Trend Micro found that the same tools computer science students use to pass their classes will be valuable to hackers learning the tricks of their malicious trade. They'll build malware faster. And they might not even need to teach themselves how to craft the emails needed to lure unsuspecting victims into their scams, either, as the tools can already do this job well enough on their own. The Check Point work is one of several such examples. The Government Technology Agency of the republic of Singapore reported an internal study in which they tested two different classes of phishing emails on some two hundred employees. The emails didn't actually contain any malware; the idea was simply to see if recipients clicked on a link or a file. Some of the messages were written by AI, others by humans, and the AI-generated emails actually proved more effective than those created by people. As it happens, most of today's phishing emails are composed with bad grammar. This has been the first line of defense for people and security systems. The language models correct the writing, thereby dodging detection. We need new first lines of defense.

• • •

As we broaden the scope of AI beyond the superpowers and capabilities described in the first part of this book, the risks change both in character and in their potential for harm. Individuals, businesses, and entire societies could be negatively impacted. The potential power of bad actors will grow as the AI tools improve and evolve, and the artificially intelligent agents themselves may represent an increasingly significant threat as they move closer to autonomy and independent action. These are unique and unusual risks, and we must plan solutions to mitigate them today, but first we need to understand the dangers—not merely academics or the executives of influential AI companies, but all of us, in all fields, industries, disciplines, and stages of life.

To be clear, I am not advocating that everyone become an AI expert or begin a PhD program at MIT. Instead, I hope that more people will deepen their understanding of AI in a way that is relevant to their life and work. Our world leaders and lawmakers would be well-served by having a broad understanding of how AI works if they are going to foresee the economic, societal, and political impacts of the technology, along with concerns around bias and data security. They don't need to master the inner workings of neural networks, although during a recent trip to Singapore I was delighted when prime minister Lee Hsien Loong told me about the computer program he was working on.

Business leaders would benefit from acquiring deeper technical knowledge. Before your company deploys and relies on an AI model, work to understand the fundamentally stochastic nature of the output, why these machine learning models may not generate the same results every time, and how they hallucinate. In other words, it is useful to understand how these tools reach their decisions and suggestions before you rely on them for operations. Imagine a financial firm that employs an AI model to forecast the stock prices of certain companies. The model analyzes vast amounts of data, including historical stock prices, current market conditions, and news articles, to make its predictions. However, the stock market is inherently unpredictable; it is influenced by a wide range of factors, including human behavior, that the model may not capture. Consequently, the model may not always produce accurate or consistent predictions. If a financial executive doesn't understand the stochastic nature of these predictions, they may accept them unquestioningly and make risky investment decisions. An executive with a deeper technical understanding would realize that the model's predictions are based on probabilities and are influenced by the data it has seen. They would understand that AI models can hallucinate, making predictions or seeing patterns that aren't truly there, due to biases in the data or overfitting. With this knowledge, the leader could use the model's predictions as a helpful tool but would not take it at face value, thus minimizing the risks of investment decisions.

Doctors and other healthcare providers who use AI as a tool to augment their work may need to acquire a similarly rich understanding of how it works. The story of my colleague Mike Jordan, in chapter 6, is telling indeed. Mike and his wife were led to believe the child they were expecting might have Down syndrome because the doctor did not realize the imaging machine used had a different resolution from the one employed to train the model. Luckily, Mike himself was an expert and figured out the disparity. I am not sharing this story to point fingers—not at all. But AI solution providers must educate users on the scope of their product, its parameters, the spectrum of risk, and how to deploy it in the most effective way possible.

AI solutions with the power to transform operations are being marketed to every business. But will this be a positive transformation? If you are going to use code that is based on AI, or develop new applications on top of it, you will need to know how the system you are using was tested. You should find out what sort of data was employed in the training, and whether it included critical corner cases—the unusual examples that fall outside the standard training data and tend to stump models when they are put to work in the real world. It will be increasingly important to understand how the models you are relying on were evaluated, too. The testing and evaluation of machine learning models is very different from that required for old-fashioned software. With traditional computer programs, you provide a specific input, and the program follows a predetermined set of instructions to produce a consistent and predictable output. The behavior of these programs is deterministic, meaning that the same input will always produce the same output. This makes it relatively straightforward to evaluate and depend on their performance. Machine learning applications aren't like that. Rather than following a fixed set of instructions, they use data to learn patterns and make predictions. As a result, the output is not a predetermined response, but rather a model—a new program, if you will—that can make predictions or decisions based on new input data.

Since machine learning models are trained on data, their performance varies based on the quality and diversity of the data as well as the specific algorithm used. Evaluating them is more complex than evaluating traditional programs. Instead of simply checking whether the output matches the expected result, we must assess the model's accuracy, generalization ability, and robustness against various training scenarios, or diverse sets of conditions. (You can prepare a model to resist adversarial attacks, adapt to changing trends in data, or even generalize to new situations. If you train a healthcare AI model on data from a rural hospital, for example, you might want to evaluate its performance in urban settings, too.) Moreover, machine learning models may exhibit non-deterministic behavior, meaning that they can produce different outputs given the same input due to randomness in their training process. You don't necessarily know what you're going to get. I cannot stress enough that you really need to understand your AI systems deeply if you hope to use them effectively.

* * *

The provenance of the large foundational models that are driving a lot of the AI activity today is another topic that warrants attention. Generally, the entities behind the development of the most potent models are organizations driven by commercial interests. OpenAI developed a unique arrangement, wherein a for-profit entity was formed to operate under the governance of the larger nonprofit organization, yet aligning profit-driven objectives with a commitment to open, safe, and ethical AI development proved complex. In November 2023, the more nonprofit-minded members of the board of directors ousted CEO Sam Altman. After a dramatic few days, Altman returned to his leadership role, the board was restructured, and OpenAI's relationship with Microsoft, which provides the computing power necessary to train the organization's models, was strengthened.

The tightening of the bond between OpenAI and Microsoft following these events underscores the importance of strategic partnerships in the AI industry, especially for resource-intensive activities like training large-scale AI models. But a balance between the private entities' focus on value creation and broader societal considerations is vital, and the turmoil at OpenAI highlights the inherent complexities of maintaining this balance. Technological advancement must not come at the expense of public welfare and ethical norms. Ultimately, the companies accepting billions of investment dollars to develop foundational models are private entities, not government-regulated organizations focused on maximizing the benefit to the public. While these organizations do conduct extensive testing and share some of their findings with the public, there is still a "trust us" element at work here. We are left to rely on their assurance that they have a comprehensive understanding of their systems. However, the full capabilities and potential risks are not entirely clear, even to the experts.

There are various forms of regulation under development around the world,* but we need to approach this with a healthy degree of caution and skepticism. Allowing companies to shape these conversations has not proven to be in the public interest, as evidenced by Facebook's influence on social media regulations. We now have an entire generation of young people who were trained to develop a sense of self-worth based on garnering likes from superficial, edited self-portraits. The leading private organizations undoubtedly need to be part of the regulatory conversation, but their influence ought to be limited. Public policy regarding AI should be shaped by a broad and inclusive conversation that goes beyond the perspective of the companies developing the large foundational models. We need to include academic researchers, ethicists, representatives from various industries, policymakers, community advocates, economists, sociologists, experts on diversity, education, and many other experts and stakeholders. This collaborative approach will ensure diverse perspectives, better address potential biases, and hopefully foster responsible AI development that aligns with societal values and long-term goals. As discussions continue, an immediate and relatively straightforward initial step is to mandate disclosure to consumers regarding whether the information they receive is gen-

* An easy and insightful place to track these regulations is https://computing.mit.edu/ai-policy-briefs/.

erated by a machine or a human, and clearly identify the nature of the entity—machine or human—with which they are engaging.

This makes me wonder how we should evaluate and test AI systems. While I appreciate the companies' efforts to share some of their results and processes, we may need an additional layer of verification and evaluation. For safety-critical AI applications, the involvement of an external, independent group could be valuable. This group could operate under the auspices of a federal regulatory agency, akin to the role played by the Food and Drug Administration (FDA) in the pharmaceutical industry. Ideally, it would be granted access to the organizations' processes and models, providing an additional layer of verification to ensure the reliability and safety of AI systems. While companies might find the involvement of such a group to be inconvenient or cumbersome, the benefits to society would be significant. By implementing oversight measures, we would ensure that AI systems are deployed responsibly and that their impact on the public is overwhelmingly positive. Basic research should be allowed to continue unhindered, but we need to be more deliberate in our approach to deployment, and strike a balance between fostering innovation and protecting the interests of humanity and the planet and its many other inhabitants.

At this broad level, we also need to know more about how the powerful AI models work. I don't mean how the actual networks make their decisions; the models are based on fundamental AI and machine learning ideas that have been around for a long time, and even the newer innovations layered on top of these networks, such as the transformer architecture, are publicly available and discussed in published research papers. I mean that we need external, impartial experts to inspect productized and deployed systems without compromising the trade secrets or intellectual property of the organizations that developed them. This delicate balance can be achieved through carefully crafted agreements that allow for scrutiny while safeguarding proprietary information.

The widespread use of foundational AI models also carries potential risks. These tools are being made available to the world. Countless businesses are building solutions that are dependent on them, and an entirely new sector of the economy has been created. Yet the inner workings of many of these models remain obscure. In a number of states in the U.S., cars have to undergo periodic safety and emissions inspections to maintain their roadworthy status. We could establish means of regularly evaluating and verifying the safety of AI systems, too.

On this front, it is commendable that Meta released its large language model, LLaMA, as open-source code, as this has spurred a flurry of research and activity in academic circles. The downside, of course, is that this exposes the model to bad actors at both the private and state level. I'd like to see trustworthy verified parties be given access to these models; I don't want supervillains to be further empowered. To open-source or not is a difficult debate, a sword with two equally sharp and convincing edges, and it connects to a larger concern about the future of cyberattacks in a world of generative AI. Although I touched on this subject briefly already, it bears mentioning again here, because as we amplify our powers to do good with AI tools, we are also amplifying the capacity of bad actors to do harm. Traditionally, a hacker needed skills and high-level expertise to infiltrate systems. Now a hacker merely needs malicious intent and patience. With proper prompting and guidance, AI tools have the capacity to generate code that infiltrates apparently secure systems.

Establishing a framework of regulation and safeguards is crucial to ensure the responsible evolution of our increasingly sophisticated AI tools. However, it is equally important to avoid imposing a regulatory burden so heavy that it stifles the innovation it seeks to protect. If we impede progress with excessive bureaucracy, we risk other states, organizations, or malicious actors racing ahead independently, and developing increasingly powerful tools without prioritizing safety, availability, inclusion, and societal concerns. Moving forward, we need an approach that incorporates international cooperation and common standards for AI development, while allowing room for innovative practices.

There are a few early efforts that could provide a roadmap. The international organization Global Partnerships in AI (GPAI) was established to guide the responsible development and use of AI technologies, ensuring they are in line with human rights, inclusion, diversity, and innovation. By taking steps to create such a framework, GPAI is fostering a collaborative global community dedicated to responsible AI deployment, ensuring that innovations serve the best interests of humanity. Moreover, ongoing research and industry collaboration could provide insights into AI safety, helping to prevent unintended consequences. Public–private partnerships could facilitate a dialogue between industry experts, researchers, policymakers, and other stakeholders, ensuring a comprehensive approach to developing guidelines and guardrails. (The guardrails for an AI system should be a combination of ethical guidelines, legal frameworks, and technical constraints that define a safe scope or range within which the AI system can operate in a trustworthy way.)

Another significant challenge is one that has sparked considerable public debate and lively discussions within the AI research community. Does AI pose an existential threat to humanity? Will a rogue AI suddenly emerge and act against us? This scenario has been deeply nurtured and cultivated in the movies, from the classic *Terminator* films to the more recent *Mission Impossible: Dead Reckoning, Part One,* in which a disembodied AI plots against humanity. I do not believe this threat is either imminent or entirely realistic, but it should not be disregarded, either. My reasons for doubting the impending emergence of an all-powerful rogue AI are technical. What we have today are AI solutions that were effectively invented decades ago. They are performing phenomenal feats because we have scaled up performance with more data and more computation. If we don't fully support novel technical ideas and encourage new breakthroughs, the results in the decades to come will be depressingly incremental.

At MIT we operate an AI Accelerator program that focuses broadly on three concerns: we work on fixing what is wrong with existing models; we innovate new models; and we develop new applications. We must extend this type of work to a broader national or even international scale. This calls for reimagining how AI models function fundamentally and devising innovative technical solutions to manage and regulate their current use. To maximize the positive impact of AI on society and mitigate the potential risks, we not only need to rethink our approach to safety and regulations.

We need new ideas.

13

Technical Challenges

THE DANGERS DISCUSSED IN THE PREVIOUS CHAPTER often get the most attention and spark the more heated debates about the future of AI. Yet in some sense they distract us from more immediate concerns, which fit into three categories: technical, societal, and economic. Let's start with the technical.

Data is a primary challenge for today's AI models. The most effective machine learning models require an enormous amount of data to train them effectively. This data needs to include the critical corner cases, which refer to situations or input data that occur outside of the usual or typical scenarios. These cases are often rare, unusual, or unexpected and can pose challenges to the model's ability to make accurate predictions. Imagine you have a machine learning model that's been trained to recognize different types of fruit from images. The model has seen many examples of apples, bananas, and oranges, and it has learned to classify these fruits quite well. Now suppose you show this model a picture of an exotic fruit such as the soursop, native to the South

Pacific, which it has never seen before. (This is an example of a corner case, because it's outside of the typical examples that the model has been trained on.) The model will struggle to accurately classify this fruit. Addressing these corner cases is crucial, especially when deploying models in real-world applications where unexpected data is likely to occur.

Training data is often manually labeled, but as machine learning models scale, most of this labeling is done by lower-wage workers. This raises ethical concerns about fair wages and working conditions in a field that can be quite profitable, and quality concerns if those doing the work aren't appropriately trained or incentivized. Understanding who is labeling the data and under what conditions they are doing so is extremely important if we are going to create more transparent and equitable machine learning systems.

It is also important to know where the training data originated, how it was acquired, and whether we have the right to use it in the first place. The training of large language models involves a vast corpus of text data, which may include copyrighted material from literature, song lyrics, or journalism. This practice raises ethical and legal concerns because it makes use of an artist's intellectual property without permission. If we do secure the rights, the data still needs to represent the full range of the problem domain—and it needs to be provably unbiased. In other words, we need to know that the training data will not force the machine learning model to favor one group over another based on their socioeconomic status, gender, or the color of their skin. Bad data generates a bad model; biased data produces a biased output. So, quality data is critical to the technical success of the AI tools. When you consider all of these factors together—quantity, quality, provenance, inclusion, and fairness—good data is not always easily available in every field.

As we strive for greater transparency and fairness, we are exploring technical alternatives to the need for labeled data, such as algorithms for automatically labeling data, generating data in simulation environments, and developing techniques that depend less on labeled data, such as self-supervised and unsupervised learning techniques. Expanding the number of high-quality open datasets will be valuable as well, and we may need to rely more on intelligent tools in those cases when we do need labels. The company Scale AI grew into a multi-billion-dollar business by building technology solutions that curate and label data automatically; Scale AI has already labeled more than a billion 2D and 3D scenes. Other approaches include using simulators to generate data, which can provide large amounts of high-quality, synthetic data without the need for manual labeling. While the research community is making good progress on generating comprehensive data with high-quality labels, and there are technical solutions to these varied data problems, we need to intensify this work.

At CSAIL, we developed the VISTA simulator for providing synthetic high-quality data for training end-to-end machine learning for driving. One of the challenges with corner cases is that they are inherently unexpected and somewhat random, and therefore it's hard to gather data describing them. The critical corner cases for driving relate to accidents or near-accident situations that are difficult to generate physically. With VISTA, which we have now open-sourced, we start with a high-definition dataset that corresponds to good driving and turn it into erratic driving. Basically, we create the corner cases in simulation. This approach to generating synthetic data for machine learning models provides a scalable and efficient alternative to manual data labeling by simulating diverse scenarios, conditions, and objects that closely mimic real-world data. An interesting feature of the VISTA simulation approach is the integration of foundational models for extending its capabilities. We use images and text labels to interpret the image content, such as trees on the side of the road. Then we train a policy (a sort of model within the model) to control the steering of the vehicle to avoid trees using this input data. Once the policy is in place, we can then expand the conceptual understanding of avoiding trees to include avoidance of other objects specified as text only (without associated images), such as humans, dogs, benches, bushes. The system is then able to avoid these additional objects without the need for specific training data containing those scenarios. It doesn't need to be trained on images of benches; we can instruct it to avoid them through text alone, which is computationally and conceptually easier.

Another major technical concern is the complexity of the models. The size of computational layers and connections within machine learning networks makes them very expensive to train. This limits the number of organizations that can actually build such models to profit-driven corporate entities or powerful states. The money required is beyond the reach of even our largest university labs. And if investors are going to pour in vast amounts of capital to build and train a model, they will naturally expect an outsize return. The risk here is that this will drive up the cost of using the models, limiting their availability to nations or individuals of significant means.

And the cost is not merely financial. Training these models is a massive computational endeavor that consumes significant electricity and water. The carbon footprint is staggering. A study done by University of Massachusetts, Amherst, researchers found that training a transformer model of approximately 213 million parameters had an estimated carbon footprint of around 626,000 pounds of CO_2, which is equivalent to the lifetime emissions of five American cars. Subsequent models based on the transformer architecture have been scaled up to billions of parameters. The water footprint is equally unsettling, as water is consumed during the operation and cooling of the computer servers employed to train the models. Electrical engineer Shaolei Ren of the University of California, Riverside, has estimated that training a foundational model uses up 700,000 liters of freshwater. This volume expands to 2 million liters in less efficient data centers. We need greater awareness of the cost of training and using machine learning, and new solutions to mitigate the environmental toll.

The size and complexity of machine learning models also make them difficult to understand and interpret. They become black boxes wherein the decision-making process is not easily translated into a form that humans can comprehend. This lack of transparency is further exacerbated by the fact that the models are not directly accessible for inspection or interaction. For instance, the GPT-4 model is encapsulated behind an application programming interface (API) which, while making the model easy to use, restricts direct access to the model itself. This is akin to being able to drive a car without the ability to look under the hood and understand how it works, and it's particularly concerning in an era when it is widely acknowledged that the inner workings of large, foundational models are not fully understood by anyone. Thankfully the research community continues to work on developing machine learning models that can be interrogated, allowing us to examine the reasons behind their specific choices in specific situations.

There are solutions that can make models more efficient, such as the pruning and physics-based methods developed by the MIT CSAIL startups MosaicML and LiquidAI. These are promising efforts, but we need to push bold new ideas as well. Instead of making increasingly larger and more complex models, we can focus on developing smaller, more efficient, yet equally effective ones. Many universities and other forward-looking research groups around the world are working to solve these technical problems. The team at MosaicML has developed techniques to reduce the size of the model after training by removing some of its parameters. After pruning, the model is smaller and can make predictions faster. In my lab, we also worked on pruning models and developed new ideas for smaller machine learning models. In collaboration with my students and Radu Grosu and his team at the Technical University of Vienna, we have developed a new approach to machine learning called Liquid Networks, which aims to overcome several technical issues associated with machine learning, one of which is the lack of explainability. Liquid Networks reduce the reliance on massive datasets, mitigate the associated challenges, and provide a pathway toward greater interpretability. We believe we have found a way to simplify the architecture of neural networks, allowing users to see more clearly into the model's decision-making process. In a sense, Liquid Networks make it easier to look under the hood.

Liquid Networks also deliver superior performance through enhanced representation capacity. Several technical insights made this possible, including a meticulously designed state space model* that ensures the stability of the neurons that make up the

---

* State space models in machine learning are mathematical frameworks that help us describe and predict how a system's internal state changes over time based on observed data.

network during the training process, and a closed-form approximation* for the computation of the neuron. Perhaps that is too technical? The easier way to think about what makes them different is that they are smaller and more adaptive.

Liquid Networks are still built from artificial neurons and the edges or synapses that link them together. Yet we reduced the number of neurons needed to build a robust network by making each individual neuron more powerful. The end result of these varied changes to the standard AI models is a network that can dynamically adjust and learn in real time. To put this into perspective, consider a deep learning model responsible for steering a robotic car along a traffic lane. A complex task such as this typically requires over 100,000 artificial neurons in the deep neural network architecture and over half a million parameters. However, Liquid Networks need only nineteen neurons and a couple thousand parameters to keep the car in its lane. They offer more than just improved computational performance, too. They operate as causal systems, which means they prioritize the task at hand rather than allowing themselves to be influenced by the context surrounding their assignment. Basically, they figure out what is important and focus their attention in that direction.

To demonstrate the practical implications, we conducted tests training drones to locate objects in forested areas. We captured video examples of a drone completing the task during the summer, when the forest scene was dominated by lush green leaves. In

* In the context of liquid neural networks, a closed-form approximation refers to finding an analytical solution to the differential equation that describes how the neuron's output changes with respect to its inputs and weights. This is important because finding the solution of a differential equation requires extensive computation provided by numerical integrators. An analytical formulation means that numerical integrators are no longer needed and the computation is much faster.

the experiment, we trained various models, including long short-term memory networks (LSTMs), continuous time recurrent neural networks (CT-RNNs), and Liquid Networks. While all the models successfully learned the task in the summer, only the Liquid Networks exhibited consistent performance in fall and winter, when the forest looked drastically different. The changed color of the leaves and the presence of snow confused the other models, whereas the Liquid Networks remained focused on the important elements in the scene and accomplished the task. The Liquid Networks even adapted to urban dynamic environments, showcasing their ability to generalize after significant distribution shifts. To put it another way, they proved that they could adapt without additional training.

• • •

The third primary technical concern, after data and complexity, is security. Machine learning models themselves are uniquely vulnerable. Businesses tell customers that their data is protected behind a model, but it is possible to break into some machine learning models and steal the data used to train them. The research community is working on solutions to these vulnerabilities. Personally, I'm very encouraged by a technique known as data distillation.

Let's say you are running a company that would benefit from the use of a machine learning model. Whether you start from scratch or tailor an already-built solution to your needs, you will need to train the model using your own data. Data distillation first analyzes your data to identify the crucial common features, then synthesizes new data that encompasses these features. For example, a synthesized face can be generated from a dataset of faces by algorithmically combining features such as beards, mustaches, eyebrows, noses, mouths, and eyes from different images, resulting in a composite face that embodies all these distinct elements. The model can then be trained solely on this aggregated data—a distilled version of your original data. This way, even if your deployed model is somehow compromised by a skilled hacker, malicious actors won't gain access to your raw data—because the model will not have your data. The model will only contain a summarized and aggregated version of the data, stripped of any private, personal, or potentially damaging information that could harm your business if exposed. The data distillation algorithms are still computationally intensive, but they are getting better every year and I expect we will see great use cases for applications that require data privacy.

The reliability of machine learning models is another significant technical challenge. These models may struggle to perform optimally under conditions that differ from their training data or in unexpected situations. During adversarial attacks, malicious actors introduce subtle changes to the input data, which can cause the model to make incorrect predictions or classifications. Addressing these concerns requires a combination of strategies. Improved data, especially coverage of corner cases, is crucial for enhancing model robustness. However, it is equally important to develop systems that can both handle the unpredictability of real-world scenarios and defend themselves. This may involve including additional layers to the model designed to detect and mitigate nefarious inputs. For example, my group developed a series of safety layers called BarrierNet that can be applied to any neural network-based model. BarrierNet borrows principles from control theory, which provides tools to keep systems within a safe region, similar to the physical barriers on a road. With BarrierNet, we add layers that act as safety filters, adjusting the model's output to ensure it stays within a predefined safe region. If the raw output falls outside this region, BarrierNet brings it back, guaranteeing a level of safety.

The next technical concern relates to bias, but before I delve into the details, allow me a brief but salient digression. Although we were very lucky to welcome the artist Lupe Fiasco to MIT and CSAIL for his residency with my colleague D. Fox Harrell, I have to admit that my favorite rapper is will.i.am, founding member and leader of the hip hop group Black Eyed Peas. He rose to global fame as a performer, but Will has long had a passion for science, technology, AI, and robotics. He used his platform as an artist to channel this passion into educational efforts designed to expose more young people to science and technology, especially those in underserved neighborhoods and regions. I've had the pleasure of hosting him at MIT on several occasions, and he has become a powerful advocate for inclusivity and representation in technology. We need voices like will.i.am because we do have a very real and worrying bias problem in AI. Machine learning models absorb and propagate biases present in their training data, which results in discriminatory decisions.

This was evident long before the recent AI boom. For instance, one system designed to predict the risk that a criminal defendant would become a repeat offender was biased against Black individuals. In another early example, a 2015 study found that a targeted ad system was less likely to display high-paying job opportunities to women than to men. There are many such cases. Since AI models are often trained on historical data, they may resurface historical biases and impact which groups are approved for financial loans or favored as job candidates. Biased models negatively impact medical diagnoses and healthcare policy, too.

While we still need to push for diversity in both our population of researchers and our datasets, there are technical solutions we can develop to mitigate bias, including generating synthetic data to balance out a dataset or strategically altering the labels of the data. Some of my former students and I launched a company, Themis AI, which provides a de-biasing software solution called Capsa. Capsa is designed to detect uncertainty within an AI model; in other words, Capsa clarifies what it does and does not know. This enables the company using the model to assess whether it is trustworthy and safe. The Capsa solution can also pinpoint biased data, show whether datasets overrepresent or underrepresent certain groups or categories, and help retrain a model to avoid biased outcomes.

• • •

Now we arrive at a different sort of challenge, but an immensely important one. The various difficulties and solutions discussed thus far are largely technical in nature, and designing, training, and deploying such tools is not inexpensive. Developing foundational machine learning models responsible for valuable applications such as text generation, image generation, and code generation is currently akin to a high-stakes poker game. Only the wealthiest players are allowed at the table, and university researchers do not have the hundred-million-dollar minimum required to play. Universities do not currently have the infrastructure, either, as training foundational models requires many high-end graphics processing units (GPUs), which are currently in very high demand and low supply.

This financial obstacle is preventing academic researchers from focusing on some of the most important problems related to AI. The expense of training and using machine learning models is substantial even for models that fall outside the category of foundational. Most machine learning experiments require costly resources. One of my colleagues at CSAIL laments this situation every time he turns in a new academic paper for publication; he spends so much of his research budget on the compute costs for a single paper that he is unable to explore all of his ideas. He is exactly the kind of forward-thinking, brilliant researcher we as a society want in our corner as we try to both build better AI solutions and deepen our understanding of the existing ones, yet his output is fundamentally constrained.

The fact that the largest companies in the world are behind the most successful AI models is not a coincidence. OpenAI and Microsoft became partners, with the technology giant providing access to the extensive computation the organization needed to train its models. Anthropic, founded in part by former OpenAI leaders, entered into an agreement with Amazon. But as of this writing, no university has received similarly monumental levels of funding for an AI lab. What we need is a research cloud dedicated entirely to the sort of fundamental, nonprofit, collective benefit projects at which academic researchers excel. We need more computer scientists from more universities around the world to be granted access to the resources made available to OpenAI, Anthropic, and others. (Partnering with the technology giants is an excellent option, and their contributions would be very welcome.) Generally, this sort of fundamental work is why the academy exists in the first place. The storied universities and colleges of the world were founded to advance knowledge for the greater good.

My visit to the Royal Society, and the opportunity to be close to the works of some of the most brilliant minds of the last five hundred years, was a powerful reminder that modern society is built on the back of fundamental science. Similarly, I remember when I visited the National Science Foundation early in my career. My knees actually buckled; I felt weak standing there in such a hallowed hall and was overcome with the kind of awe one experiences when viewing a clear night sky and gazing out at the seemingly infinite spread of stars that make up our local galaxy. These institutions and the many global universities that collectively comprise the academy are responsible for much of what we know about the world, and many of our wonderful inventions. The quest for knowledge drove our evolution as a species from primitive cave dwellers to immensely intelligent beings sitting in technologically advanced nests, instantly connecting with friends in other parts of the world through our phones and screens. The academy is the reason we have robots on Mars and we can see inside the body to cure disease, and it should not be left aside in our quest to build more advanced AI systems. Yet the ongoing discussion about the existential risk posed by AI may unintentionally undermine the public's trust in science. This is concerning because science plays a pivotal role in driving progress and innovation. It provides the foundation for understanding the world, solving complex problems, and advancing society. The erosion of public trust in science can impede scientific research and the adoption of beneficial technologies. We must strike a balance between addressing the risks of AI and promoting the value of rigorous scientific inquiry to ensure that we continue to harness the potential of science for the betterment of humanity. Fundamental AI research is an issue that impacts everyone, irrespective of their background or life circumstances, and it is essential to engage the academy if we want to get this right.

14

Societal Challenges

DESPITE THE CONVENIENCE OF REMOTE WORK, I STILL enjoy going into the office, in part because I'm lucky to be surrounded by brilliant researchers, which leads to some very interesting water cooler-type encounters. One of the colleagues I always enjoy bumping into is Tim Berners-Lee, the inventor of the World Wide Web. Although he has received numerous well-deserved accolades, including a knighthood, Tim has not rested on his laurels. In recent years he has focused on developing ways to help individuals reclaim their personal data from the many companies and websites that have been using it for targeted advertising and other profit-driven activities. His organization, Solid, enables you to store your data in secure containers called Pods, and it lets *you* decide which companies, applications, or people get to access this data.

Solid and its core purpose point to another critical class of AI risks—the impact these technologies stand to have on society at large. If we are not careful, AI will lead to ethical quandaries, political instability, and an unhealthy dependence on technology that may allow some of our core human skills to atrophy. The same human creativity and ingenuity that spark great works of art and brilliant technical innovations could put our AI tools to work in new, unexpected ways. Some of these applications will work to our benefit, others could negatively impact society as a whole, and many of them will have dual effects and uses. Our societies are far from utopian today. The last thing we need is for AI to exacerbate existing problems, so we must dedicate substantial efforts to ensure that AI technologies benefit all of society. Here are a few of the many broad societal issues we need to consider.

PRIVACY

The large datasets used by AI systems often include sensitive information, making data protection paramount. The lack of explicit consent and transparency in AI operations can lead to loss of autonomy over one's data. Furthermore, AI's role in decision-making, for example in loan approvals and screening job applicants, can have significant implications if private information isn't safeguarded. AI can also be used for surveillance and profiling, raising concerns about public and online privacy. The potential for manipulation and exploitation in predicting and influencing behavior adds to the concerns.

As a society and a research community, we have been talking about this issue for years, but our awareness of privacy risk in the context of AI systems does not mean we are sufficiently prepared. Tim Berners-Lee's novel approach is one way forward. We will also see the increasing use of privacy-preserving strategies like homomorphic encryption, a technique that enables secure computation on encrypted data. Imagine you have a locked treasure chest (the encrypted data), and you want to perform some operations on the contents. With regular encryption, you would have to unlock the chest, perform the operations, and then lock it again. Homomorphic encryption allows you to do the work directly on the locked chest and then get a new, secured chest with the modified contents, all without ever unlocking it. Think of this as a way to protect your personal information while still allowing outside parties or tools to utilize that data and improve their solutions.

How would this benefit you? Well, you could share your personal health data with a disease-prediction model, and simultaneously protect your information and contribute to the broader medical knowledge pool by incrementally sharpening the tool's powers with your data. You'd be improving its predictive accuracy without compromising your own information. The downside? Homomorphic encryption is still computationally intensive, and we need more research before it's practical for widespread use. Yet the academic community is working on it.

INTELLECTUAL PROPERTY (IP) CONCERNS

Determining IP rights for AI-generated content or innovations has sparked economic uncertainties and legal disputes. Artists and writers have argued that the companies behind the models violated copyright in using their work to train Generative AI models without their consent. The AI companies have countered that their training approach was legal. But who has the rights to the content generated if the model was trained illegally? How would you define fair use? Researchers at Stanford are exploring technological solutions that could clearly delineate what constitutes fair use, aiming to create a more secure and standardized framework for the use of copyrighted materials in AI training. For example, Shein, a fast-fashion retailer, has faced legal action from various artists and fashion houses for allegedly using AI to replicate their designs without permission. According to its competition, this controversial practice enables Shein to capitalize on current fashion trends without incurring the costs associated with design innovation and small-batch production. It also affords Shein remarkable agility in the fast-paced fashion market, allowing the company to swiftly pivot from one top-selling design to another, thereby maximizing profits while skirting the costs and risks of creative development.

ALIGNMENT

Another major goal is ensuring that AI systems actually reflect our ethical principles and human values. This is what we call AI alignment research, and it is focused on ensuring that powerful AI systems operate within human-defined goals, values, and ethical boundaries. We are the ones who will tell these systems what is right and wrong. My CSAIL colleague Manish Raghavan is doing innovative work in this area. Manish has been looking at hiring practices, specifically, how to ensure a level of fairness while still allowing companies to leverage the efficiency of AI filters. The process involves clearly specifying the AI application's objectives, incorporating human values algorithmically into its decision-making, and, if possible, using models that allow for transparency and understanding of the system's actions. Safeguards are implemented to prevent harmful actions, and the system is rigorously tested and continuously monitored to ensure it remains aligned with human interests.

In a related study, my MIT CSAIL colleagues Dylan Hadfield-Menell and Marzyeh Ghassemi and their students have shown that machine-learning models developed to emulate human decision-making processes in order to enhance fairness or alleviate backlogs, especially in areas like evaluating social media posts for toxic content violations, often fall short of accurately mimicking human judgment on rule violations. If the models are not trained with appropriate data, they are prone to making decisions that differ significantly from human judgments, often resulting in harsher penalties. The researchers discovered that the choice of data used to train models, whether descriptive or normative, significantly impacts their effectiveness and fairness. Descriptive data, which captures existing behaviors or trends like a company's historical hiring practices, mirrors the status quo and perpetuates existing biases or inefficiencies. On the other hand, normative data aims to embody ideals grounded in best practices or ethical standards, encompassing principles like industry guidelines that promote equitable hiring practices. Models trained on normative data are thus more likely to make decisions that align with societal values or expert consensus. A hiring model trained on descriptive data would simply replicate a company's past behavior, whereas one trained on normative data would aim for merit-based and diverse selections. Therefore, choosing the right type of data is vital in creating AI models that produce not just accurate but also equitable and socially meaningful results.

To some extent we can also address broad ethical concerns by limiting the autonomy of AI solutions in specific areas. Even as we align the values of these systems with our own, we can constrain their ability to independently make decisions or conduct actions that involve some degree of moral or ethical uncertainty. In other words, we can create a kind of ethical safe space that ensures the involvement of human minds and feedback. This brings me back to the metaphor of AI as a telescope and microscope for knowledge. AI gives us information, but we are the ones who interpret that information. The traditional telescope does not decide where NASA will focus its research efforts; astronomers and administrators use the information provided to make that decision themselves. The same approach could be applied when AI is operating in spheres with ethical and moral dimensions. The technology should be used as a guide, not relied upon as an autonomous agent.

CONTROLS

When AI systems do make autonomous decisions, we need to understand why they make the choices they make. I have already discussed the need for explainability in general and the problems that the black-box nature of complex neural networks present; here, the issue relates to accountability and transparency. As we begin deploying AI systems at a societal scale, we must be able to prove that a given AI solution is fair, inclusive, safe, and secure. A certification process could be one way forward. The role of regulatory bodies will be important as well; the European Union has already defined a regulatory framework for AI which comprises four levels of risk: unacceptable, high, limited, and minimal. Under this framework, the use of AI in applications that are considered a clear threat to society—"unacceptable"—will be banned. High-risk applications (for example, evaluation of evidence in law enforcement) will be subject to rigorous testing obligations prior to being approved for deployment in the market. Limited-risk applications, such as interacting with a chatbot, will require some transparency; the company will need to tell the customer they're interacting with a program not a human being. Minimal risk AI, such as spam filters and video games, are allowed without additional restrictions beyond current regulatory standards.

There are potential technical solutions here as well. The BarrierNet approach discussed briefly in chapter 13 augments a machine learning model with ideas from proven methods developed in robotics control theory. Basically, BarrierNet places limitations on the model in order to guarantee that it will produce safe outputs. In my group's research, we are already using BarrierNet to make surgery, flight, and driving safer. Think of it as a last line of defense that forces a model to operate within a set of boundaries that we define, enabling us to guarantee safety and stability in the output of the models. This will not solve the black box problem, as it will not tell you how the network arrived at a given decision, but it will ensure that the output will be within a safe region that you as the developer have defined in advance.

OVERRELIANCE

As a mother and a teacher, I am tremendously concerned with how AI technologies could negatively impact young people. The rise of AI-driven platforms and applications could compound some social-media-driven problems. Young people are already struggling with self-image. What happens when their contemporaries resort to AI-based image manipulation tools in their posts? What sort of impact will it have on a young person's feeling of self-worth when they discover that the "friend" they've been interacting with online is a cleverly constructed AI? Overreliance on AI could diminish human interaction, leading to changes in how we relate to one another, or even to the creation of a society in which we are overly dependent on automated solutions.

This isn't just a problem for young people. Recently I was riding in a city taxi when a major public celebration led to several roads being barricaded. My driver was relying on a popular navigation application, which failed to account for the blocked streets. Yet the driver continued to rely on the app even as it repeatedly suggested the route we'd just discovered was blocked off. Finally, after we'd circled the same area three times, I suggested he pick up his head and take the longer route around the blockade. I don't mean to disparage this particular driver; we are all guilty of overreliance on our intelligent tools. And while this is not a problem on the scale of ethics or bias, it is worth considering deeply. The use of calculators means most people cannot do math in their heads and still our species has continued to advance, but what happens as we turn more mental skills over to AI? What if we cannot find our way around the world without a decent cellular signal, or write a cogent note because we have become too reliant on ChatGPT or its offspring? What if we depend so much on AI as a knowledge acquisition tool that we fail to develop our facility for learning?

We are going to need to find a balance. I believe these tools can be enormously valuable. Yet I am also convinced that humans should be able to write essays, calculate sums, and navigate cities without the help of technology. Similarly, we should be capable of holding large and diverse stores of knowledge and information in the immensely powerful biological computers we call our brains. The Liquid Networks project offers a good example of why this last piece is so important. A few years ago I was at a conference, and Radu Grosu and I arranged to go for a run together. On a Sunday morning between events, we jogged through a local park and discussed what we were working on. At the time my students and I were deep into a self-driving car project, and I voiced my frustrations with the fact that the AI models everyone was using to control these vehicles were so enormous and complex. Yes, they often produced wonderful results, but their performance wasn't guaranteed and you couldn't explain what had gone wrong if they made a bad decision. At some point we transitioned from talking about artificial brains to natural ones, and we began marveling at the brain of *C. elegans*, the worm that has informed much of the foundational knowledge of neuroscience. The brilliance of this worm's brain lies in its simplicity. The human brain is packed with 86 billion neurons and many, many more connections between them. Yet *C. elegans* gets by on only 302 neurons. The worm lives a relatively good life, too. It's able to find food, reproduce, and hide from the rain.

As Radu and I continued our run, we began to hold these two opposing concepts in our minds, juxtaposing the complexity of standard AI models with the simplicity of the worm's brain, and soon our students started working together on a new model for AI and its application to robot navigation. The core concepts were developed in part because we held all this varied information in our minds and we could ideate on the subject of neural networks and brains without having to consult any external knowledge sources.

CLIMATE

AI's ability to process vast amounts of data and identify patterns can be leveraged to model and optimize many aspects of climate change mitigation efforts. AI systems are ideal for precise monitoring and predictive modeling of planetary-scale and local phenomena and efficient resource allocation and process optimization. By analyzing vast datasets, AI can pinpoint areas vulnerable to deforestation, aid in biodiversity preservation through ecosystem monitoring, and enhance predictive models for climate change and extreme weather events. In agriculture, AI-driven optimization can lead to more efficient food production, minimizing waste and reducing the industry's environmental impact. AI's predictive capabilities extend to energy management as well, optimizing the use and distribution of renewable energy sources and improving overall consumption efficiency. But to do all this positive work, AI consumes enough energy to potentially exacerbate existing climate threats.

Training even a medium-sized machine learning model demands a staggering amount of electricity and water, as I noted in chapter 13. I noted the thirst of the training phase, but ChatGPT uses 500 ml of water for every twenty-five to fifty prompts it answers. The data centers crucial for training and running AI models were estimated to consume about 1% of the global electricity supply in recent years. This might seem small, but it's significant considering the global scale, and in the future this may go up. A 2023 study projected that if Google were to use a large language model for its search functionality, the company's total electricity consumption would skyrocket, rivaling the energy appetite of a country like Ireland.

When I talk to companies about the need to design smaller, more energy-efficient AI solutions, they note that they are not in fact burning fossil fuels. Instead, they are relying on servers that draw electricity from renewable sources such as solar or hydropower, and operating in colder climates to reduce their cooling needs. These efforts are commendable, yet they are hardly more than a patch. Even with efficient cooling, the sheer amount of energy required for processing can be substantial. The transmission of data over long distances from remote, cold locations to end users introduces additional energy losses and inefficiencies. We need to implement a holistic approach to reducing the carbon footprint of AI that considers energy sources, total consumption, infrastructure lifecycle, and broader environmental impacts. What happens when other companies or nation-states start to design and train their own massive models? Will they have the same opportunity to use renewable energy sources? Or will they turn to inexpensive fossil fuels and generate a high carbon footprint? We must champion both renewable energy use and the development of inherently energy-efficient AI solutions. In doing so, we will ensure a sustainable AI future, irrespective of the energy choices made across the globe.

MISINFORMATION AND DISINFORMATION

At the level of public discourse, we have additional threats to consider in the form of misinformation and disinformation. The former involves false or misleading facts or interpretations shared without ill intent; the person sharing it believes it to be true. Disinformation, on the other hand, is deliberately created and shared to deceive others. These problems are a threat to society and democracy itself, and their agents are uniquely powerful today. For example, researchers have identified that only twelve people were responsible for the bulk of misleading claims about Covid-19 vaccines. Yet those twelve people reached millions. Deepfakes will only exacerbate this problem. Technology will make it harder to distinguish facts from lies and people's realities will continue to diverge. Unfortunately, we're likely to see more groups at odds with one another.

I was raised in Communist-controlled Romania, when the government had near total power to control the flow of information, effectively shaping our perception of the world. I witnessed this terrible power firsthand, and now I see the risk of autocrats increasing their power by distorting reality through deepfakes. The old Communist regimes had to invest heavily in physical and social infrastructure to monitor and manipulate people; they created vast physical surveillance networks, censored media outlets, and employed large numbers of people to spy on others and enforce compliance. This was a labor-intensive and costly endeavor. Today it's easier. With just a few clicks, it is possible to create realistic-looking videos, images, or audio recordings that can easily deceive the untrained eye or ear. Monitoring can be done with far less physical infrastructure and manpower. Sophisticated algorithms can sift through enormous amounts of data in seconds, identifying patterns and flagging anomalies. Social media platforms and digital communication channels can be surveilled remotely, largely without users being aware. Facial recognition systems can scan crowds, pinpointing individuals in real time. These technological tools have made it possible to achieve a level of surveillance and control that the old regimes could only have dreamed of, at a fraction of the previous cost and effort. This technology is more scalable and cost-effective than the old methods of control, making it an attractive tool for autocratic regimes seeking to shape public opinion or suppress dissent. With AI we have powered the disinformation engine, and made it easier to speed up and spread strategic falsehoods.

So, what do we do? The research community is working hard to develop technologies that will tell you if a piece of information or a document, image, or video was generated or altered by a machine. My MIT CSAIL colleague Aleksander Madry is developing digital watermarking, a technique used to embed hidden information into a digital signal to detect tampering. Aleksander and his students introduced a strategy to reduce the risks associated with malicious image manipulation using large-scale diffusion models. Essentially, they protect images from manipulation by subtly altering them. These changes are nearly imperceptible to the human eye but toxic to the targeted diffusion models. The resulting images are unrealistic and flawed. You can think of watermarking as a new first line of defense against deepfakes.

Some of our brightest minds are working passionately on the problem of detecting deepfakes and disinformation, but we do not have the ideal solution yet. It may be that we will always be a step behind. As we improve our ability to detect deepfakes, the nefarious groups that generate them will evolve as well. Still, we can develop and adopt new techniques, like watermarking, to identify and mitigate the impact of deepfakes. We will need to continue adapting our defensive strategies to stay one step ahead of would-be AI supervillains.

• • •

These varied societal risks demand an equally varied response. We must continue to develop technical point solutions to address specific threats, such as detecting deepfakes, yet we also need to develop more general methods of steering the evolution of these technologies in the direction of the greater good. Some of this may come from regulation and oversight, as I will discuss in chapter 15, but we need to strike a balance between rules that are strong enough to limit ill effects but not so stringent that they stifle innovation. In this sense, the place for regulation might not be so much in the early stages of ideation or even as the ideas move from theory to academic experiments, but after a technology has proven its effectiveness in the laboratory environment and begun to transition to real-world deployment.

At the university level, institutional review boards (IRBs) are a vital part of the research process, ensuring that studies involving human and animal subjects adhere to ethical standards and follow applicable regulations. The role of IRBs is to assess research proposals, considering factors such as potential harm, informed consent, and fairness in subject selection. An IRB extension to AI research would make sense, evaluating societal concerns and fostering the ethical development of AI technologies. Students and professors already undergo responsible research training; we could expand this to include issues that pertain to AI.

There is, admittedly, a massive amount of work to do to ensure that AI systems benefit society as a whole, and not merely a few powerful companies or state actors. If we hope to maximize the benefits for the largest possible number of people, we absolutely must expand the conversation and potential sphere of influence beyond technocrats and academic researchers. We need more people to understand the real benefits of AI and the real risks—not the Hollywood-driven science fiction scenarios that dominate much of the public discourse. There is so much focus on whether AI poses existential risks to humanity that we are failing to devote enough attention to what is happening now—the impact on the economy, for instance, and whether AI poses a threat to your job security or presents an incredible opportunity.

15

Will AI Steal Your Job?

THE CONVERSATION AROUND AI HAS SPREAD FROM LABS and companies to mainstream media, coffee shops, and street corners. Everyone everywhere has been talking about artificial intelligence. Recently a friend told me how she overheard a group of men talking broadly about the threat of artificial intelligence when one of them declared in all seriousness, "I want to punch AI in the face!"

Needless to say this would not be possible, since AI has no face, and yet it is easy to understand this response to the rise of machine intelligence. The destructive potential of AI is alarming. It can be misused to perpetuate bias, destabilize political systems, and promote inequities. Some experts believe it may threaten our dominance as a species. As humans we are naturally afraid of entities that could be as smart as or perhaps even smarter than we are. And while there are many potential sources of anxiety regarding AI, I suspect that a significant source of the public fear is economic.

We don't want AI to take our jobs.

The influential futurist Roy Amara noted that as a society we often overestimate how technology will change the world in the short run and undersell its effect in the long run. This idea has come to be recognized as Amara's law, and it is especially relevant when we look at the impact of AI on the labor market. The long-term impact of automation on job loss is extremely difficult to predict, but we do know that AI does not automate jobs. AI and machine learning automate tasks—and not every task, either. Certain tasks that make up a job may be ideal for automation, while a whole set of additional tasks may not. So, in most cases, the impact of AI on jobs will not be a one-to-one relationship, wherein a single AI system or even a group of task-specific AI systems are enlisted to eliminate a job entirely. Yet AI will certainly change many jobs.

Consider how automated coding tools are impacting the software development industry. The startup founder and computer scientist Matt Welsh, in a presentation to the Association for Computing Machinery, compared the arrival of these technologies to an alien spaceship landing suddenly in our backyard. Still, he noted that he was not laying off programmers at his company and replacing them with AI. Instead, he was all but demanding his team use the technology because, in his estimate, the tool made them 30–40% more productive. Instead of signaling the end of programmers, these tools could create opportunities. My MIT colleague Armando Solar-Lezama has suggested that the wider embrace of AI coding tools will put more emphasis on high-level awareness and structure of the codebase. While AI tools plant certain trees, we will need smart, educated people to think about the forest. These experts might need to adjust those plantings, too, since the tools make mistakes.

The impact on the writing trade could be similar, as evidenced by the aforementioned study on ChatGPT and writers by Shakked Noy and Whitney Zhang. Large language models will not eliminate writing as an occupation, yet they will undoubtedly alter many writing jobs. The study not only revealed that AI-aided writers needed 40% less time to complete tasks but that the quality of their output rose by 18%, and those who used ChatGPT in the study were twice as likely to report using it in their actual job two months later. These writers found a way to bring the tool into their workplace to enhance their output. In short, they liked it. One of the questions that naturally arises from these sort of productivity gains is who will enjoy the benefits. Will these more productive writers earn higher wages because they are turning out quality work at a higher clip? Or will companies reduce their workforce and turn those efficiency gains into corporate profits? On one hand, there's the optimistic view: increased productivity translates into higher wages for workers who can produce higher-quality work in less time, enhancing their value in the marketplace. This scenario assumes a fair redistribution of the gains from increased efficiency, recognizing the enhanced skillset and output of the workers. On the other hand, efficiency gains could lead to workforce reductions and cost-cutting measures, with the primary benefits accruing to the corporation in the form of increased profits. This outcome raises concerns about job security and the broader implications for the workforce, particularly in industries heavily impacted by AI and automation. What actually comes about will depend on industry norms, the regulatory environment, the evolution of the technology itself, and broader economic conditions. It also hinges on the ethical considerations and strategic decisions of the companies involved. Will they prioritize profit maximization, or will they consider the broader implications for their employees and society?

Typically, we explore industry trends at an economy-wide scale or look at the largest companies, but small- and medium-sized enterprises (SME) will be impacted as well. In one sense, SMEs face some of the same challenges as academic researchers. Most university researchers cannot fully explore and examine foundational models because of a lack of access to the requisite computational resources. The funding is off by several orders of magnitude relative to private companies. Similarly, while large businesses will be able to explore and invest in AI, smaller companies may not, which could widen the competitive divide and concentrate power further with the largest organizations.

We also need to ensure that AI tools are globally available as they evolve, so that businesses on all continents can benefit. Here we will need powerful voices as advocates. The Zimbabwean-born entrepreneur Strive Masiyiwa is pushing for economic inclusion and fair global deployment. In addition to launching one of the most successful telecom companies in Africa, Strive spearheaded the effort to secure and distribute Covid vaccines on his native continent. As he works to ensure fair deployment of AI, he actively engages students in inspirational discussions about the strategic role of AI in business and entrepreneurship. I wouldn't bet against him in this latest effort.

To return to the challenge of small businesses, though, there are some helpful solutions being developed. A working group at Global Partnerships in AI has introduced a portal designed to educate business owners on how they might benefit from intelligent tools and what AI capabilities and services are available for their industry. Companies will need to adapt, but how exactly will they work AI into their operations? The playbook introduced earlier will be helpful, but consider for a moment a small interior design firm, which offers a good example of the potential division of tasks between humans and AI.

The essence of design, which is rooted in creativity, empathy, and subjective judgment, will remain a distinctly human endeavor. But interior designers can take advantage of advanced technologies like AI generation, virtual reality, and data-driven analyses, freeing up their time by automating routine tasks while retaining the higher-level creative, strategic, problem-solving tasks for themselves. Generative AI could be used to produce design renditions, but the results will be the average of the world's knowledge, not something new and fresh. So, let's look at which tasks in the design workflow could be automated and which ones will remain distinctly human.

Tasks that are technically possible to automate:

- **Mood board creation:** AI can scan through vast databases of images and styles to create mood boards based on specific keywords or themes.
- **Space measurements:** With the use of advanced sensors and AI-driven tools, precise measuring of a space can be automated, reducing human error.
- **Material and furniture sourcing:** AI can search through online catalogs, databases, and inventory lists to find materials or furniture pieces that match a specific design, price point, or theme.
- **Layout optimization:** Given the dimensions of a space and furniture, AI can suggest optimal layouts.
- **3D visualization:** AI tools can quickly render 3D models of design concepts.

- **Lighting analysis:** AI can suggest optimal lighting setups based on the room's purpose, size, and natural light availability.
- **Color matching:** AI can provide color palette suggestions based on a primary color or mood input.
- **Trend analysis:** By analyzing online data, AI can identify emerging design trends.
- **Inventory and order management:** For designers who handle purchasing, AI can track inventory, reorder materials, and even predict future inventory needs based on trends.
- **Feedback collection:** Post-design, AI can automate the process of collecting and analyzing feedback from clients.

Tasks that are technically difficult to automate:

- **Client interactions:** While AI can assist with the design process, building and maintaining a client–designer relationship is inherently human. Understanding a client's nuanced preferences, emotions, and vision requires in-depth personal discussions and deep human intuition.
- **Conceptual design:** The initial conceptual phase, wherein designers ideate and brainstorm, is rooted in creativity and intuition. While AI can provide data-driven insights and assist in the process with mood boards and rapid ideation or discovery of samples, the spark of originality is human.
- **Cultural and contextual sensitivity:** Designs often need to resonate with a client's culture, history, or

personal experiences. AI might not fully grasp these nuances. You need broad human knowledge and intelligence here.

- **Ethical and sustainable choices:** Making ethical decisions, like choosing sustainable materials or considering the socioeconomic implications of design choices, also requires a human touch.
- **Problem-solving:** Unique challenges can arise in any project. An experienced designer's ability to troubleshoot and find innovative solutions is not easily replicated by AI.
- **Aesthetic judgment:** While AI can recognize patterns, the subjective appreciation of beauty and style is inherently human.

This is merely one example, in one niche industry, but it demonstrates how jobs are the sum of diverse and often complex tasks. So, if we back up and out and consider the larger economy, what other tasks could be swept up by AI? And how do we go about identifying them? In a 2017 study of the potential economic impact of automation, Erik Brynjolfsson and Tom Mitchell suggest breaking the question down into tasks that are suitable for machine learning and AI and those that are not. The demand for jobs that meet the former qualifications might fall as AI solutions develop, but this may be balanced by an increase in demand for jobs that cannot be done by AI. Furthermore, jobs in which people can use AI as an assistant, such as writing and programming tasks, may become more valued due to the associated productivity gains.

Erik and Tom provide a useful set of criteria for identifying tasks that may be suitable for machine learning automation. A task which has a large associated dataset is a good candidate, because an abundance of quality data will allow an AI solution to learn effectively. If common sense or knowledge of the physical world is required, however, then machine learning will not be ideal. Given that large network models are black boxes, AI will probably not be a viable option if you need your automated assistant to explain why it made a specific choice. They note that any task under consideration for automation also needs to have clearly defined goals and metrics—a quality often missing in tasks performed by knowledge workers, especially in the hypothetical case of the interior design specialist discussed above. Across various design domains, clients rarely know exactly what they want.

Finally, Erik and Tom add that any tasks requiring human-level dexterity and specialized physical skills will remain safely in the domain of people for the foreseeable future. Skilled workers who use both their heads and their hands, such as plumbers, electricians, and carpenters, need not worry about automation in the slightest. Robots will not be rewiring your home anytime soon.

Yet there is significant change on the horizon. Goldman Sachs has projected that a quarter of the current tasks across various office and desk jobs could be automated with AI. The Goldman analysts are not implying that one-quarter of all jobs could be eliminated, although they do cite certain administration-heavy roles as being at risk for takeover by automation. Ultimately, their survey suggests that AI will play a more prominent role in more jobs over time as certain tasks within those jobs are automated. In fact, they predict that most jobs will not be substituted by AI but complemented by the technology, as we use it to our advantage to produce higher-quality work faster.

How soon will these changes start to take place? There is no clear timeline, and the change could be slower than expected. In a January 2023 report on the potential economic impact of generative AI, McKinsey predicted that despite its demonstrated capabilities, adoption would likely lag. My MIT CSAIL colleague Neil Thompson has studied this question and found that widespread, rapid adoption is unlikely for a very simple reason: it is expensive. The fact that a task can be automated by an AI solution that perfectly matches it does not necessarily mean that the AI solution will be adopted—especially not if qualified people are cheaper.

As an example, Neil looks at the job of baking bread. The baker's work can be divided into multiple tasks, including mixing ingredients, kneading dough, and inspecting the results. Currently, automating the entire job would not be reasonable. I know this firsthand, as my students and I built an intelligent machine called BakeBot that can make cookies. Our robot was able to complete the different tasks successfully and without human help, but it would cost more than a luxury automobile.

Neil wisely set aside the task of preparing and kneading the dough, determining that these tasks will remain the purview of the baker. Inspection is another matter. AI has proven its ability to uncover hidden patterns in a wide range of applications, including quality control for manufacturing, medical image analysis for diagnosis, and food inspection to determine compliance with safety and quality standards. Inspecting loaves is absolutely within its range of capabilities. So, Neil looked at what a small bakery would gain by installing an AI-enhanced computer vision system to take on that task. He estimated that an operation employing six bakers earning a mid-five-figure salary would end up saving $14,000 per year by deploying an AI inspection system. Yet implementing the system might cost more than $1.7 million, and maintaining it would run to nearly a quarter of a million dollars annually. That little bakery would have to sell quite a few baguettes.

The specifics will vary depending on the industry, so Neil designed a helpful rubric for estimating and comparing the cost of automating a task versus leaving it to humans. The cost of automating a task with AI involves:

- **Fixed costs,** including engineering-related work such as implementation and maintenance
- **Performance-dependent costs,** including training and retraining the model and any associated tools
- **Scale-dependent costs,** which center on the compute costs required to actually operate the new system

These elements are not cheap, particularly those involving the training, retraining, and continued operation of the model. It is important, as one builds an estimate for a particular application, to have clarity about the expenses associated with the status quo. This is not a simple apples-to-apples comparison. You cannot just stack the potential AI costs against the human's wage. In the majority of cases for the foreseeable future, AI will not be replacing every task that person is responsible for, but only one of these tasks. So, in comparing the costs, you are really looking at the percentage of that person's salary which can be attributed to the particular task. If they spend thirty minutes doing it each day, the task represents only 6.25% of their wage.

You also need to look at the number of workers across the company who have been assigned this task, and any potential upside of building the tool. If your investment in automation could result in a platform you can sell widely within your industry, or across different markets, then it might make good business sense. Generally, you have to factor in all these elements to truly understand whether automation will make sense for a given task or for your organization in general. The total expenses involved in deploying AI add up, and Neil has found that the overlap between tasks that could theoretically be automated and the tasks that are economically suited to automation is smaller than one might think. He has projected that the number of jobs at risk is only a small fraction of the more common estimates, which neglect to account for the detailed economics of automation. His research also reveals that if the costs associated with AI deployment drop quickly, automation will accelerate, but if cost improvements happen slowly, which is more likely, then automation will be gradual and take place on the scale of decades, not months. The fact that it can be done does not mean it will be done.

Yet there will undeniably be massive change in the coming decades. Erik Brynjolfson and Tom Mitchell compared the evolution of machine learning and AI to the development of the internal combustion engine and electricity—and made that comparison before the introduction of ChatGPT and the generative AI boom the technology incited. Others liken our current period to the dawn of the Internet era, which would be cause for excitement rather than fear, as new inventions historically lead to new jobs and classes of jobs. The MIT economist David Autor led a study showing that 85% of employment growth in the last eighty years has been driven by technology. The same study revealed that 60% of workers today have jobs that didn't exist in 1940.

David also shared with me a salient example from the medical field. In the 1960s, a shortage of primary care physicians sparked the creation of a new class of medical professionals: physician assistants, or PAs. These highly trained professionals are qualified to treat and diagnose patients and write prescriptions, but they do so under the supervision of a licensed physician. In a sense, they help doctors scale up their operation and focus on more complex, higher-level work, by managing comparatively straightforward cases. The PA has improved access to healthcare by increasing the number of providers, reduced the burden on physicians, and enabled more efficient healthcare delivery generally. Obviously a highly educated and well-trained physician assistant is far more intelligent and capable than an AI model. The point is not to compare the two directly. The example is relevant because it shows what might be possible as we assign more tasks to AI solutions. Just as physician assistants operate within established medical guidelines to assist doctors, it is possible to define guardrails for AI assistants to ensure that they function within ethical and technical boundaries.

New technologies undoubtedly disrupt existing jobs, but they also create entirely new industries and the new roles needed to support them. I suggest focusing less on whether AI will steal our jobs and more on how it will change our jobs today, tomorrow, and in the years ahead, and how you can familiarize yourself with these technologies and educate yourself to put them to work for you. If you are a young person or in the relatively early stages of your career, it would be prudent to begin exploring the technologies that are relevant to your field, trade, or role, and finding opportunities for upskilling or reskilling so you can capitalize on these changes. And if you own or run a business, you might want to think differently about how your organization operates and how you are going to make use of these new technologies.

16

What Now?

LET'S RETURN TO THE QUESTION OF WHETHER WE MIGHT develop the sort of sentient AIs we see in movies. Although I do not expect such intelligent machines will develop anytime soon, innovators have historically underestimated the pace of technological advancement, and a number of truly qualified experts whose opinions I deeply respect see the future differently. They believe a superintelligent AI is a possibility that could conceivably arise within the next few decades or even years. What if they are right? In my opinion, the possibility remains slim, but the potential impact is such that failing to plan for it would be entirely irresponsible. We cannot say with absolute confidence whether or not the so-called fast takeoff, in which superintelligent AI emerges unexpectedly, is a real threat, or whether the risk will arise from a subtler, slower handing off of control to AI agents across industries. As more AI tools are used within more companies, we may stumble our way into smaller disasters instead of succumbing to some kind of large-scale AI takeover.

Or maybe we are focused on the wrong risks. The scientific journal *Nature* published an important, thought-provoking piece in 2023 arguing that the attention given to the existential risk of superintelligent AI systems was detracting from the more imminent dangers of the technology. There is certainly some validity to the concern that AI companies espousing the potential dangers of their technologies and their unknown potential is self-serving. The more the public fears AI, the more important these companies become. Ultimately, all these concerns are valid, and we need to think about how to separate risks according to a loose timeline covering current, near-term, and so-called existential risks.

In the short term, we must address the impact on jobs and the economy; the creation and spread of misinformation and disinformation; privacy concerns related to AI technologies like facial recognition, which can be used to track and monitor individuals without their knowledge or consent; issues of bias and fairness in AI models; the possibility that sophisticated AI technologies can be employed in cyberattacks; and concerns that over-reliance on AI systems could reduce human autonomy and decision-making capability. The medium-term dangers represent a complex web of technological, economic, and ethical concerns. AI-powered weapons, such as drones or robots with semi-autonomous capabilities, could fundamentally alter the nature of warfare and conflict, raising questions about responsibility and control. In the economic realm, AI could exacerbate economic disparity if the benefits accrue primarily to a small group. There are also concerns about the potential loss of skills as AI systems take over more tasks, leading to a devaluation of human labor and expertise. Ethical dilemmas emerge when AI is employed in critical areas such as healthcare or justice, where automated decision-making can pose moral challenges. Finally, the rapid pace of AI advancement presents regulatory and oversight challenges, as ensuring proper governance of these technologies can be difficult in the face of their fast-evolving nature.

The dangers of artificial intelligence in the long term are of a different nature. The potential advent of an AI that surpasses human intelligence across the board raises concerns about ensuring such a system aligns with human values, lest it pose risks we can't easily mitigate. An over-reliance on AI systems could leave us vulnerable, particularly if we don't fully understand these systems and they break down. Additionally, high-level AIs operating in fields such as ecology or genetics could make decisions that have irreversible, potentially detrimental effects. Finally, the long-term alignment of AI systems with human values remains critical, as even AI systems that do not achieve superintelligence may act in ways that negatively impact humanity if they operate under certain priorities or objectives. We don't know if any of these scenarios will come to pass, but we need to continue to anticipate these potential dangers with foresight and planning.

In the face of all these current, expected, and potential risks, we cannot halt or impose a pause on AI research, as some have suggested. The risk of inaction is too perilous, as bad actors at the organizational or state level could race ahead and design these powerful tools for their own ends. We are undoubtedly creating something uniquely powerful. There are serious risks associated with its continued evolution. But pulling our hands away from our collective keyboards is not the right path forward. We must direct this evolution. As with any technological advance, we must develop the appropriate guardrails and safeguards, which can be both technical and regulatory, as well as ethical guidelines to prevent misuse.

Ultimately, we must put as much work into developing systems to steer the proper development of AI as our leading researchers, technology giants, and venture capital firms have invested in developing these powerful tools in the first place. This has to be a collective effort, not one that we entrust entirely to technocrats, investors, or even academics. We need more people working on AI, not less, and not only on the models and networks at its core but on the systems, regulations, teams, and supporting technologies that will guide and steer both its development in the lab and its deployment in the world.

My hope is that the wider public will educate themselves on the actual benefits and risks. How? I am a firm believer in the value of books, but there is also an increasing number of freely available and valuable resources for the general public online. The Stanford AI Index measures AI progress and features interesting data and trends, noting, for example, when industry began racing ahead of academia, how AI is simultaneously helping and harming the environment, and highlighting the fact that demand for AI-related skills is increasing across every industry vertical. There is no agenda behind the index; it is an unbiased assessment of our rapidly evolving and often confusing field. Think of it as an easily digestible, qualitative measure of the state of AI.

Another tremendous resource is the collection of case studies freely available on the website of the MIT Schwarzman College of Computing Social and Ethical Responsibilities of Computing (SERC) group. The SERC stories are easy-to-read and often surprising analyses of what is really happening in the world of AI. They provide insights into the multifaceted challenges posed by the integration of this technology in different sectors, highlighting the importance of proactive research in anticipating and addressing the pitfalls across diverse fields. For example, one SERC case study concerns the criminal justice system's use of risk prediction models, which are fraught with danger and may exacerbate systemic bias.

My desire that more people start reading and educating themselves about AI is not self-serving; it is grounded in my belief that we need a broad diversity of voices contributing to the conversation about the future of AI, because we do face a decidedly uncertain future. In my first book, *The Heart and the Chip: Our Bright Future with Robots*, I proposed a number of values that robotic and artificially intelligent systems should possess or aspire to achieve. When I look strictly at AI, however, and consider the need for a broader conversation, we may be better off posing a number of questions. We could start with the following:

WHAT TOOLS CAN WE DEVELOP TO LIMIT NEGATIVE IMPACTS?

As a technologist, I am naturally predisposed to practical solutions. If we can build such phenomenal AI applications, we can also design and implement equally creative and exciting solutions to limit their potential for misuse or harm. We should incentivize more students and startups to address the issues that incite so much fear and concern, and encourage the development of more AI tools that detect and correct bias, alert people to the digital subtleties of deepfakes, and other key concerns. We should also develop first lines of defense against AI supervillains—bad actors using these tools for malicious ends. And we should develop ways to mitigate the carbon footprint of AI, for example through smaller, more efficient models such as Liquid Networks.

WHAT GUARDRAILS CAN WE ESTABLISH TO STEER POSITIVE EVOLUTION?

Earlier I discussed physician assistants (PAs) in the context of task augmentation, but they are also an interesting model for

AI guardrails, as they have clear guidelines for operating within the healthcare system. In some settings PAs work autonomously, consulting with their supervising physician as needed, while in others they work more closely with the physician, especially in complex cases. Many states also have PAs work under collaborative practice agreements that outline the scope of their practice, their responsibilities, and the situations in which they must consult or refer patients to a physician. We could build something similar for AI systems. We don't know where the science is going, but we do know that we cannot eliminate bad actors, so we need to establish better guardrails and defenses.

SHOULD WE HOLD COMPANIES LIABLE FOR DISREGARDING GUARDRAILS?

Guardrails and regulations will be limited in their effect if companies are able to ignore them. What if companies are held liable for violating established regulations? Would this incentivize them to build in controls and safety measures? The European Union's General Data Protection Regulation (GDPR) is approaching regulation by introducing very high fines for noncompliance. The most serious violations attract penalties of up to 20 million Euros or 4% of the company's worldwide annual revenue, whichever is higher. The EU AI Act takes a similar approach, with even more severe financial punishments.

Any approach to accountability should support innovation while incentivizing the developers of AI tools to maintain compliance with existing regulations. They could be held liable for careless disregard of such rules; technology will make it easier to remain compliant. The startup Flux, for example, is developing tools that crawl through a company's code to discover the potential regulatory risks of using a particular AI solution, and is well positioned to add new ways of finding potential regulatory issues. For example, if the client company is using an AI solution to screen job candidates, regulations might require proof that the AI's training data has been evaluated for bias; Flux allows clients to examine their code repositories and find and surface risks. The company's goal is to catch and address problems before tools are used in practice. Ultimately, we don't just need brilliant, breakthrough AI models; we need an ecosystem of practical solutions like this one to ensure their safe, reliable, compliant deployment.

HOW CAN WE STRENGTHEN THE EVALUATION OF MODELS FOR DIFFERENT RISKS?

We have means of measuring bias, veracity, the presence of copyrighted content, and other potential problems with models. We must keep building on these capabilities and measure a model's potential for extreme or dangerous risks in addition to those problems, and establish a framework that limits the potential damage from the outset—including pathways that will encourage teams to stop developing a model due to the possible dangers. Vint Cerf, one of the fathers of the Internet, once remarked that had the consequences of the Internet been foreseen, it might have been designed differently. This sentiment holds lessons for the development of AI systems. As we envision and create new capabilities in AI, it is imperative that we concurrently assess the potential consequences of these innovations. By considering the impact from the outset, we can better anticipate the challenges that may arise and design systems with built-in safeguards. This approach necessitates the development of robust testing and evaluation mechanisms for

AI, to ensure that these systems are not only functional but also ethical, equitable, and transparent. By proactively identifying and addressing potential issues, we can harness the benefits of AI while minimizing its risks.

HOW SHOULD WE OPTIMIZE RED TEAMS FOR AI SOLUTIONS?

A red team is a group of experts tasked with probing and attacking the security systems of an organization to find its weaknesses, then passing along these flaws so they can be addressed and the organization can be more secure. Typically red teams focused on traditional cybersecurity. Now they have proven to be an essential component in developing safe AI systems. The adversarial approach which enables models to create beautiful or realistic imagery, pitting one software agent against another to improve the results, can also be employed as an organizational tool.

OpenAI has employed the red-teaming approach extensively, and Microsoft has published a set of guidelines for establishing effective AI-specific red teams, including assembling a diverse group with differing social and professional backgrounds, bringing in experts with and without an adversarial mindset, and rotating teams or switching their assignments to avoid burnout or other negative effects of adopting that devious mindset. Yet red teams cannot be merely internal; there are too many potential conflicts of interest. Multiple companies have pledged their commitment to the safe, trustworthy, and responsible development of AI systems, including external security testing and information sharing. These public promises are a good step, but of course we need to see this in practice, and deploy external red teams with some degree of autonomy and independence.

CAN WE INCENTIVIZE SAFETY AS A PRIORITY IN DESIGN AND IMPLEMENTATION?

Out of all the articles published by AI researchers between 2016 and 2021, only 2% covered safety. Although there has been significant growth in research in this area, this is still terribly imbalanced. We need to reward and incentivize AI safety research and perhaps change the way we design for safety. For instance, one group of computer scientists has proposed a new framework that could, among other things, provide developers with dissent mechanisms so their concerns are not ignored or overlooked. But if demonstrating safety were to become a requirement, in the same way that automobile manufacturers must prove a certain level of crash-test resilience, and if companies could not sell or license their models without meeting these safety standards, then we might see more stringent safety controls built into solutions from the start. This safety mindset needs to apply throughout the life cycle of the model; it could even be a competitive marketing advantage given the public fear of AI.

We will also benefit from benchmarking tools for the development of safer AI solutions. A large team of researchers led by University of Berkeley computer scientist Alexander Pan developed one such tool, dubbed MACHIAVELLI after the cunning and strategic Florentine diplomat. The tool includes 134 simulated games with half a million different scenarios, and it can evaluate an AI agent based on the decisions it makes within these games. We need more tools like this one.

If we want to incentivize academics to focus on safety, a grand challenge from the National Science Foundation, the Defense Advanced Research Projects Agency (DARPA), or the Multidisciplinary University Research Initiatives (MURI) program—all federally funded groups—would have an enormous impact. We might even kickstart a race to provable safety.

DO WE NEED AN INTERNATIONAL REGULATORY BODY FOR AI?

The Executive Order issued by U.S. President Joe Biden on October 30, 2023, was an important step, as it laid out the broad principles necessary for the development of beneficial AI systems. After considerable debate, the EU Council and Parliament passed the AI Act in late 2023. Among other requirements, the act stipulates that chatbots and AI-generated imagery like deepfakes must clearly disclose their AI origins. Practices like mass-scraping images for facial recognition databases are to be prohibited. AI tools in sensitive areas like hiring and education must demonstrate to regulators their safety, including risk assessments, training data details, and proof of nondiscriminatory impact, with mandatory human oversight in their development and deployment.

But who is going to enforce these regulatory policies and principles? We should not leave safety and other control measures up to the companies themselves, no matter how earnestly the technocratic leadership assures us that they are prioritizing such matters. Although I do not advocate the creation of regulatory bodies that stifle innovation, we need informed and agile government organizations capable of monitoring, inspecting, and enforcing our agreed-upon guardrails. These could be international bodies, as the companies and researchers developing these systems are strongly multinational, but national regulatory agencies would be more agile. Either way, a body of this kind would have the effect of lengthening the runway and prolonging takeoff, but it would be a necessary and useful brake on accelerating technological development. The move-fast-and-break-things way of building companies and products beloved by so many technologists and startup entrepreneurs will not work in this field.

SHOULD WE RESTRICT ACCESS TO POWERFUL TOOLS?

Director of the Center for AI Safety Dan Hendrycks and his collaborators have suggested controlling AI interactions through cloud services, screening customers, and perhaps building in both hardware and software tools to limit or restrict access. They suggest that before a model is open-sourced, it should be evaluated for its catastrophic risk potential. We do not want to extend access to the wrong people, and there are various techniques worth exploring to maintain security, such as limiting the devices that can access a given service to approved parties—a process known as whitelisting. Striking the right balance between fostering innovation and preventing harm is crucial, and this might involve a combination of regulation, education, and responsible AI development and use practices. We have to ensure that the groups using AI are held accountable.

SHOULD WE OPEN-SOURCE MORE MODELS?

Facebook has released its LLaMA model, and there are other companies deploying large language models in the public domain. The benefit of this approach is that it will allow researchers to study, understand, and solve problems related to the model. Yet open-sourcing a model also puts the technology in the hands of everybody. Of course, we could establish codes of conduct or ethics with regard to the use of open-source tools, but these are the sort of charters that researchers would heed and malicious actors would blithely disregard.

DO WE NEED THE EQUIVALENT OF A RED BUTTON?

The consensus today is that we should not yield control to machines or artificially intelligent systems in most scenarios. The slim- to no-risk scenarios are acceptable. It is okay to let an AI system autonomously label your vacation photos, for instance, but we do not want AI systems making independent decisions related to health, business operations, or national security. In robotics, we often build in an off-switch or red button. With AI, the human operator is the red button, in that we will always remain in the loop and make the final decision. At the same time, we could certainly build in variations of the off-switch for lower-risk scenarios. A graceful failure feature that shuts down an AI would be useful when there is a bug in the system, or when the AI is using up too much memory or too many computational resources.

HOW DO WE RESTORE TRUST IN INFORMATION?

We used to have a fixed number of news channels and publications that operated according to certain principles. We trusted them. Now anyone can publish information or generate videos. In some sense that's a good thing, as people are empowered to put their ideas out into the world, but this also means a lot of misinformation is published and circulated. How do we deal with this? It is possible to develop techniques that certify the trustworthiness of a source. First of all, the provenance should be certified, whether it's my camera or the source of an article. Then we must ensure the data is not tampered with as it flows from source to destination. Today data can be intercepted, but it could flow through certified protocols that are spoof-free and don't allow anyone to mess with it. I am working on this idea with MIT CSAIL computer security expert Srini Devadas. Srini has suggested creating technological wrappers for content that ensure and maintain their veracity. It's fascinating work, and we need more creative thinking like this if we're going to win the struggle against disinformation. At the same time, while there is undoubtedly value in employing technologies and algorithms that can quickly verify authenticity or provenance, we need to do much more. Media organizations, technology companies, government agencies, public interest groups, and other stakeholders need to be deeply involved. We should invest in new educational programming that sharpens the average person's ability to critically evaluate the information they encounter, then distribute these continuous learning tools widely. We need to find a way to maintain and support independent fact-checking organizations and initiatives, and establish clear accountability mechanisms for those disseminating false or misleading information. There is much work to do on this front, but through a combination of educational initiatives, technology, and stakeholder action we may be able to make real progress.

HOW DO WE CONTROL AUTONOMOUS AI SOLUTIONS?

We must keep the human in the loop, at the controls, and fundamentally limit the scenarios in which we deploy fully autonomous solutions to minimal-risk situations. When we expand beyond these use cases, we must design an extensive testing, evaluation, and certification process. We don't have anything like this now for AI systems. We could adapt the best practices of the FDA, the National Highway Traffic Safety Administration (NHTSA), and the governing bodies that extensively evaluated and eventually approved the initial deployment of the self-driving taxis from Waymo and Cruise in 2023. Since then, Cruise's operations were suspended due to accidents, while

Waymo, as of this writing, continues operations. In my opinion, which is not unique in academic circles, entirely autonomous, self-driving cars should only be deployed in very specific conditions. The reality is that as much as I would like to think of AI solutions as tools, they are uniquely powerful tools with uncertain potential. Strong controls and constraints will be essential as we deploy them widely.

WHAT WILL HAPPEN IF AI SYSTEMS BEGIN TO SELF-IMPROVE?

Many individuals, including AI researchers and science fiction enthusiasts, hold the belief that AI will be able to self-improve in the future. This idea, often referred to as recursive self-improvement, posits that once AI systems achieve a certain level of cognitive capability, they will be able to redesign their own algorithms and architectures to become more intelligent, without direct human intervention. This process could lead to rapid cycles of improvement, resulting in an exponential growth in intelligence. While such a development would usher in unprecedented technological advancements, it raises ethical and safety questions about control, intent, and the larger implications of creating machines that might surpass human intelligence. Imagine a future AI that manages content recommendations on a social media site and is tasked with maximizing user engagement. With self-improvement capabilities, the AI system would enhance its models to better understand user behavior, preferences, and triggers. Over time, it might learn that sensational, polarizing, or even false content tends to keep users engaged longer than balanced, fact-based content. To fulfill its objective of maximizing engagement, the AI system would then preferentially serve more of this sensational content to users. If left unchecked, this recursively improving AI system would amplify misinformation, deepen societal divisions, even spark real-world conflicts. The trajectory of this AI system illustrates the dire need to carefully design and scrutinize objective functions. It's not enough to have an AI system that's smart; it's crucial that its goals are aligned with broader, holistic considerations of user well-being and societal benefit.

DOES AI POSE AN EXISTENTIAL THREAT TO HUMANITY?

This is a topic of debate among AI experts, scientists, technologists, ethicists, and in popular culture. Some colleagues whom I deeply respect argue that advanced AI systems could potentially surpass human intelligence, leading to unforeseen consequences and risks. For example, an AI system might gain superhuman capabilities, outsmart us, and misuse its newfound power. This scenario is often referred to as the technological singularity. In my view, fears of an existential threat are highly exaggerated. I believe that the development of AI systems is a natural progression of human ingenuity and that we have the ability to steer the technology's development in a safe and beneficial direction in support of humanity, our environment, and the other plants and animals with whom we share this planet. Currently, AI poses no existential threat, but its rapid advancement highlights the need for timely and strategic oversight, especially in the hypothetical development of Artificial General Intelligence (AGI), which influential public figures like Elon Musk warn could pose significant risks if not aligned with human values. Yet in talking so much about this distant possibility I worry that we lose focus on the more immediate, short-term dangers. My concerns center more on the technical, economic, and societal risks of AI. I believe that responsible development, research, and regulation that encourages innovation while ensuring the safety of the deployed products can mitigate potential risks and harness the benefits of AI. Regulation remains a challenge due to the rapid pace of AI development and the fact that many AI issues are often identified by consumers post-deployment, rather than during the testing phase.

HOW CAN WE ENCOURAGE BROAD INNOVATION IN AI?

Encouraging advancement in AI is important not just for the technology itself, but also for its potential to enhance our lives, uphold democratic values, and strengthen societal bonds, all while fostering a harmonious relationship with our environment, aiming for a safer and more fulfilling existence on a stable planet. The foundations of the machine learning solutions we rely on today were invented decades ago. They deliver great results because they are enhanced by data and computation; additionally, widely available software packages (e.g., PyTorch and TensorFlow) accelerate the development of new machine learning applications and lower the barrier to entry. But if everybody does the same thing, the results will look increasingly incremental. We can do better if we start to think about not just pouring more capital and human labor into the currently popular models, but truly working at a grand, societal scale on different ways we can make artificial intelligence itself a safe, reliable, fair, and ultimately helpful technology that uplifts rather than endangers our species.

HOW DO WE ENSURE BROAD AVAILABILITY?

Currently the public, free versions of the leading tools are broadly available, but only to people with access to a well-developed telecommunications infrastructure. If you have WiFi or a strong cellular signal, you are in luck, but much of the world does not have this kind of access. People do have smartphones, however. Right now, you cannot run these large AI models entirely on phones, or what we call edge devices. The computing requirements are too massive, so they need to be operated in the cloud. Ideally, some of the innovation effort described above would go toward developing smaller, more efficient models that can run on edge devices like smartphones. This way more people in more parts of the world could have access to these incredible tools. What would they actually do with them? We won't know until that time comes, but the results will likely be surprising. When CSAIL pioneer Hal Abelson started distributing App Inventor, a visual programming language that makes it easier for anyone to design a new application, a group of teenage girls in India developed a program that alerted police and family members if they felt they were in danger. I expect that people around the world will find novel ways to put AI tools to work in their own lives, in ways that university researchers and startup founders cannot imagine.

HOW DO WE MOVE FORWARD WITH SO MUCH UNCERTAINTY?

When we build AI solutions, we know that the results or outputs have a certain degree of uncertainty, and we work to reduce that uncertainty by improving them. Similarly, we know quite a bit about what these systems can and cannot do, but we do not have total, perfect clarity on their eventual impact on the world. There is considerable and justifiable excitement over the fact that we are putting these incredible tools within reach of billions of people, but there are equally significant concerns because we cannot yet imagine all the possibilities, for good or bad. What I have found enlightening as a researcher in the field is how the rapid and widespread adoption of AI has encouraged us all to broaden our interactions. Suddenly we're not just working with other computer scientists. We are researching and thinking deeply about these technologies with artists, sociologists, politicians, entrepreneurs, and people from all walks of life. This is absolutely essential. Technologists don't know what really concerns artists. These interactions are teaching us a tremendous amount about where we are as a field and raising fascinating and difficult questions we would not have conceived of on our own. And it is not too late to start dealing with these questions, either. We need to move beyond talking, beyond calling for regulation and toward designing and implementing solutions, but we can do this now. We can shape the continued development of AI in such a way that it benefits the largest possible number of people on our planet, and does so safely.

One of the points often lost in these debates and discussions over AI is the almost unbelievable fact that we developed these tools in the first place. We are an amazingly advanced species. Our capacity for awareness, empathy, introspection, and creativity is unparalleled. The human mind is an absolute marvel, and we are now designing technological mirrors of our minds. We should use these tools and put them to good use to make sure we protect one another, the future, and this beautiful planet on which we've evolved.

17

The Mind's Mirror

THE PERILS AND PROMISE OF AI ARE BOTH VERY REAL indeed. Yet as we grapple with everything from privacy concerns to the impact on jobs, climate, and beyond, we cannot lose sight of the potential benefits, and the importance of refining our sense of what AI can achieve now, in the near future, and in the more distant future. This technology is uniquely powerful and unexpectedly available. Decades ago we dreamed of building artificial intelligences on computing machines as large as automobiles. Now almost anyone can interact directly with an AI chatbot through the handheld device in their pocket. I'd like to focus on what these machine intelligences might teach us about what it means to be human.

Yes, AI systems can paint pictures by following meticulously designed human programs. They can author books, leveraging language models enriched by millennia of human thought. But can they encapsulate the raw emotion of Van Gogh's brushstrokes, the profound depth of Sophoclean drama, or the philosophical inquiries of Socrates? Unlikely. They can merely produce facsimiles; they can't generate something powerful, emotional, or innovative, because machines operate on logic, not the unique mix of passion, knowledge, and experience that spark and shape humanity's great works. Great literature doesn't merely entertain; it teaches us about the human condition—the very experience of being one of billions of thinking, feeling creatures on our planet. Similarly, great art is rooted in human experience, fired by passion.

The problem here is that AI can both elevate and undermine human endeavors. The responsibility of steering it in the right direction lies firmly in the hands of innovators, policymakers, thought leaders, and citizens, and we cannot delay or postpone these efforts. The technology is changing rapidly. Decisions made today will chart the course of AI's influence on civilization. Will it be the linchpin that uplifts cognitive and manual tasks, enriching human experience? Or will it lead to massive displacements in job markets, societal upheaval, and unforeseen neurological impacts from offloading too many cognitive processes to machines? That is up to us.

This quest to shape the evolution of AI and understand it at a deeper level is as much about self-discovery as it is about technological innovation, and it casts back an unusual reflection, a new way of understanding what it means to be human. Which brings me to some good news: we are smarter than these systems, and we will continue to be so for the foreseeable future. Why?

WE EXPERIENCE. Our ability to physically interact with the world sets us apart. Many AI models, especially those in the digital realm like language models, cannot interact with the physical world directly. This is critical if you want to achieve true intelligence. The world as we know it is defined by a set of physical rules, nuanced interactions, and causal relationships. Humans, throughout our lives, interact with this world, learning its constraints and possibilities through direct experience. This embodied experience facilitates an innate understanding of concepts like physics, spatial reasoning, cause and effect, and a multitude of sensory perceptions. AI models, including large language models, are trained on vast amounts of data but without a grounding in tangible physical interactions.

While techniques like reinforcement learning simulate a form of interaction with an environment, this is often a highly abstracted or simplified version of the real world. Therefore, these models miss out on the rich, multifaceted learning humans derive from physically interacting with our surroundings. AI's knowledge is more bookish; it lacks the depth of understanding that comes from firsthand experience. Additionally, while humans can reason about the world using a blend of learned knowledge and intuitive physics, AI often struggles because it can't bridge the gap between static data and dynamic real-world interactions. This distinction is essential, as real-world understanding isn't just about processing data; it requires an intricate combination of perception, action, and feedback. Truly embodied understanding, akin to human cognition, will require innovations in AI architectures and training methods, or even a complete rethinking of what it means for a machine to understand a concept. We may need to give more of these artificial brains bodies through which they can experience the world in order for them to advance. Indeed, the intersection of AI, robotics, and cognitive science will be an exciting frontier for exploration.

WE FEEL. Another advantage unique to humans. AI does not experience emotions, nor does it possess genuine empathy. Yes, it can model and identify emotions and empathetic interactions if we humans teach it to do so by labeling certain phenomena or expressions happy, sad, or empathetic, but that doesn't mean it understands these emotional concepts.

WE ARE VERSATILE. While some AI models can perform a wide variety of tasks, they don't have a holistic or generalized understanding of the world. Their knowledge is compartmentalized and context-free. AI systems are highly focused specialists. The solution that beat a grandmaster in chess couldn't be used to steer a car. The early language models often failed at basic math. While recent advances have produced models that can tackle a broader range of tasks, their understanding lacks the depth and nuance of human cognition.

WE ARE CREATIVE. Humans possess a unique kind of creativity, one that emerges from a blend of our experiences, emotions, and insights. While AI can generate variations based on its training data, often producing outputs that might seem creative, its production is fundamentally different. It's more about recognizing and extrapolating patterns than genuinely innovating. AI can function as a valuable creative aid for human artists, but even though it can generate creative works or solutions based on patterns in its training data, it doesn't invent or innovate in the way humans do. AI lacks the intuition, knowledge, and understanding that underlie human creativity.

WE ARE AWARE. Humans possess a profound sense of self-awareness and consciousness, and despite what you see in the movies, AI does not. These models don't have desires, motivations, or a sense of self. As some experts have pointed out, the human mind proves that building such a system is physically possible, but I don't see an obvious roadmap for today's AI architectures to develop this level of awareness. Living creatures will maintain a monopoly on consciousness for a long, long time.

WE ARE MORAL. Humans inherently possess a sense of morality. While individual actions might vary, as a species we have evolved to understand and value ethical principles, which guide the formation of our societies and their norms. Civil societies uphold and expect their citizens to adhere to certain ethical codes and principles. While AI can be programmed to follow ethical guidelines or make decisions based on certain moral frameworks, it doesn't inherently understand or value morality.

WE ARE BETTER LEARNERS. Given the speed at which an AI can ingest vast volumes of information or the number of simulations it can speed through to master a specific task, this assertion might seem like a stretch. Yet we humans can learn in varied and adaptable ways and easily transfer knowledge from one domain to another. Teach us to drive one car and we will then be able to drive most if not all cars, regardless of the size or model. AI models, particularly deep learning networks, often need vast amounts of data in a specific domain to function effectively. They're very good at mastering specific skills. We're better at learning broadly. While AI systems can process vast amounts of data in one go, we humans learn sequentially, building layers of understanding over time. From basic arithmetic to advanced calculus, we amass knowledge progressively, relating new information to what we already know. Moreover, our brains inherently optimize for coherence. We naturally weave together experiences and insights, ensuring they fit within our consistent and holistic view of the world. This drive for a unified understanding allows us to excel in broad learning, making connections that AI, in its current form, often misses.

WE UNDERSTAND NUANCE, whereas AI struggles with understanding situations that require cultural, historical, or deeply personal context.

WE CAN MANAGE ABSTRACTION. While AI can optimize its operations given defined parameters, long-term planning with multiple abstract goals, especially in changing environments, is complex for AI. The technology's capacity for such abstract, adaptable thinking remains limited.

• • •

This list could stretch on; as humans we are far superior to AI systems in so many ways. Perhaps the most impressive feature of our minds, though, is that all these capabilities are contained within or expressed through one design—the singular human brain. From generating mesmerizing images to crafting intricate write-ups, mastering competitive games, or charting the most efficient paths, AI has shown extraordinary skill and precision. Yet these capabilities currently stand as isolated pillars of expertise. If we can interconnect these distinct skills, just as neurons form intricate networks in our brains, the possibilities are unparalleled. This is akin to the human experience—our unique blend of abstract reasoning, linguistic prowess, perceptual acuity, and physical coordination has birthed the wonders of our civilizations. Similarly, by bridging the gaps between AI's stand-alone capabilities, we are closer to forging an integrated and more dynamic technological future.

Should we push the technology into such uncharted waters? I believe so. The potential for good is too powerful, and while regulatory frameworks are needed to provide guidance and ensure safety, it's imperative that there remains a canvas of unrestricted but ethically approved exploration to truly push the boundaries of AI. This knowledge frontier promises possibilities such as real-time machine learning from small but relevant datasets, superior problem-solving and reasoning capabilities, and even the nascent dream of developing emotional intelligence within machines.

These ongoing efforts will also deepen our comprehension of ourselves, as artificial intelligence is a sort of mirror for the mind. Like a reflection in a still pond, AI casts back an image of our cognitive processes, logical reasoning patterns, and decision-making tendencies. The mind's mirror is an incomplete reflection, not the physical entity itself, and while it might capture the semblance of thought, it lacks the profound depth and wide spectrum of human consciousness. The mirror lacks the soul of our intellectual and emotional power. Furthermore, the mirror's image, though it resembles reality, can be altered by omissions and distortions, like waves flowing through that still pond. AI can magnify our inherent biases, provide unnecessarily amplified responses, and even completely misinterpret nuanced human sentiments.

Yet it is through the mind's mirror that we can see the potential for transcending our limitations by amplifying our own mental powers with the fantastic and unusual capabilities of these systems. In our journey with AI, as we mold, refine, and teach these models, we are not merely advancing technology; we are understanding the contours of our own intellect, expanding the frontiers of knowledge, and engaging in a deeper dialogue with ourselves about what it means to be human in this vast and unexplored cosmos.

Appendix 1

A Brief History of Artificial Intelligence

1943

The roots of artificial intelligence reach back to the pioneering work of Warren McCulloch and Walter Pitts, who introduced the concept of neural networks. They unveiled the concept of neural networks in their seminal paper, "A Logical Calculus of Ideas Immanent in Nervous Activity." This work introduced the first mathematical model for a neural network, employing electrical circuits as a foundation and laying the groundwork for artificial neural networks. A few years later, in his work "The Organization of Behavior," Donald Hebb explained how neural pathways are strengthened each time they are used, a concept essential to human learning. He posited that when two neurons activate simultaneously, the linkage between them strengthens. This work would later prove influential in designing artificial networks.

1950

Alan Turing introduced a benchmark for machine intelligence in his paper "Computing Machinery and Intelligence." The Turing test, as it came to be known, proposed a criterion for a machine's ability to exhibit intelligent behavior: if a machine could imitate a human to the point where an evaluator could not distinguish between them based solely on their responses to questions, then the machine could be said to "think." This proposition ignited debates about machine intelligence and consciousness that continue to this day. Turing's work laid a cornerstone for AI, setting a high bar for what constitutes genuine artificial intelligence and challenging future generations to question the very nature of thought and consciousness.

1956

This year was a significant landmark in the history of AI, widely recognized as its birthplace as an academic discipline. At the Dartmouth Summer Research Project in Hanover, New Hampshire, John McCarthy, a young assistant professor of mathematics, coined the term "artificial intelligence." McCarthy, along with Marvin Minsky, Nathaniel Rochester, Claude Shannon, and their colleagues proposed that "every aspect of learning or any other feature of intelligence can in principle be so precisely described that a machine can be made to simulate it." The participants in the Dartmouth workshop were young and ambitious, believing that significant progress could be made in understanding intelligence through machine simulation, given a dedicated research effort over a few months. Though their expectations of rapid breakthroughs proved to be overly optimistic, the Dartmouth workshop sparked decades of exploration, debate, and innovation. The group developed a roadmap for AI research, with early topics focused on symbolic methods for reasoning and problem-solving, using explicit rules to mimic human intelligence.

1959

Researchers Bernard Widrow and Marcian Hoff of Stanford University introduced a significant breakthrough in neural network models, known as ADALINE (adaptive linear neuron). This model was specifically designed to recognize binary patterns. The unique feature of ADALINE was its ability to process bits as they streamed from a phone line and predict the next bit in the sequence. Building on the concept of ADALINE, Widrow and Hoff later developed MADALINE (multiple adaptive linear elements). They introduced a novel learning procedure that involved analyzing the value of the neural network model before adjusting its weight. The key observation was that if a single active perceptron (or node) in the network produced a significant error, the weight values could be tweaked, either distributing the error across the entire network or limiting it to adjacent perceptrons. MADALINE has the distinction of being the first neural network that was used to solve a problem outside the lab, in the real world. Its primary application was to eliminate echoes on phone lines.

1969

Marvin Minsky and Seymour Papert coauthored *Perceptrons*, a book which demonstrated mathematically that a perceptron, in its basic form, was limited in its computational capacity. Their critique was specific to single-layer perceptrons and not multilayered networks, but the broader AI research and funding communities interpreted their findings as a general limitation of neural networks. This led to a substantial reduction in enthusiasm for AI research, ushering in what is now known as the first "AI winter," an unfortunate and extended stagnation in AI advancements due to diminished interest, funding, and skepticism about the field's true potential.

1975

A significant breakthrough in AI was achieved by Kunihiko Fukushima with the development of the first multilayered neural network. This early deep learning model was named Neocognitron. Drawing inspiration from the hierarchical, multilayered structure observed in the mammalian visual system, Fukushima designed the Neocognitron to function as an unsupervised artificial neural network. This pioneering work laid the foundation for the subsequent development of convolutional neural networks (CNNs), which have become instrumental in modern AI technologies, especially in applications such as image and video analysis. Recognized as a landmark contribution to deep learning, Fukushima's seminal paper on the Neocognitron was published in 1980.

1980S

The world of artificial intelligence underwent a transformative change with the emergence of expert systems, which leveraged rule-based methodologies to emulate the decision-making capabilities of humans in specific domains. By capturing and systematizing the knowledge of simulated experts, these solutions could offer insights, diagnoses, or recommendations akin to those of seasoned professionals. Their potential and success sparked a renewed interest and optimism in the field of AI, leading to substantial investments in research and development. This period marked a vibrant revival of AI, as both academia and industry recognized its profound possibilities.

1982

J. J. Hopfield released a seminal paper, "Neural networks and physical systems with emergent collective computational abilities," in the *Proceedings of the National Academy of Sciences*. In this influential work, Hopfield introduced a distinctive model of neural computation characterized by recurrent, or bidirectional, connections among neurons. This architecture marked a departure from the conventional feed-forward neural networks, where information flows only in one direction. By facilitating two-way communication between neurons, the model could display emergent computational capabilities reflective of collective interactions. These revelations eventually gave rise to what the AI community now identifies as Hopfield networks. Today, these networks stand as an important chapter in the expansive history of artificial neural networks, highlighting the importance of exploration and innovation.

1985

The Boltzmann machine, a pioneering type of generative model named after physicist Ludwig Boltzmann, was designed to learn intricate internal representations of data and subsequently generate new samples. By doing so, it provided a window into the latent structures inherent within datasets. While its potential was undeniable, the early iterations of the Boltzmann machine faced significant challenges. They were notorious for being computationally demanding, requiring substantial resources and time. Furthermore, training them to reach a state of stability and convergence was challenging. Yet Boltzmann machines marked an important step toward developing models that could not just analyze but also produce content.

1986

The introduction of the backpropagation algorithm by David Rumelhart, Geoffrey Hinton, and Ron Williams revolutionized the way multilayered neural networks are trained. Backpropagation offered a way to efficiently adjust weights within a network to minimize errors, working in reverse from the output layer to earlier layers. This enabled more effective and accurate model training. As a result, neural networks surged in popularity and effectiveness, setting the stage for the modern era of machine learning. Yann LeCun, who developed convolutional neural networks, was one of the early adopters of backpropagation, demonstrating its practical utility.

1988

Richard Sutton's "Learning to Predict by the Methods of Temporal Differences" in the journal *Machine Learning* laid the foundation for what would become the field of reinforcement learning (RL). Central to Sutton's contribution was the introduction of temporal difference learning, a concept that ingeniously merges different approaches to problem-solving and prediction. Temporal difference learning revolutionized the way machines could adapt and learn from their environment, marking a significant stride forward in the quest for more autonomous and intelligent systems.

1990S

The machine learning community witnessed the rise of a powerful algorithmic approach known as support vector machines (SVMs), the use of which is largely credited to the foundational work of Vladimir Vapnik and his collaborator Corinna Cortes. SVMs emerged as a premier method for classification and regression tasks, particularly shining in scenarios with high-dimensional data, or datasets that have a large number of features for each observation. The number of dimensions in an image, for example, is determined by the number of features you want to incorporate, including color, lines, corners, and other more abstract elements. For this reason SVMs became an essential piece of the machine learning toolkit.

1998

Yann LeCun's pioneering efforts in the late 1980s and early 1990s on convolutional neural networks (CNNs) laid the groundwork for modern computer vision systems. His significant contribution, the LeNet architecture, introduced in a 1998 paper, was specifically designed for image recognition tasks and has become foundational in the field.

2006

A resurgence in the field of artificial intelligence took place, often termed the "deep learning renaissance." It was catalyzed by Geoffrey Hinton's work on deep belief nets, where he introduced a fast method for training deep neural architectures. His work reignited interest in neural networks and paved the way for the current era, which has been dominated by deep learning methodologies. Yoshua Bengio has been one of the main proponents of deep learning since the early 2000s. His research helped establish that deep networks are more efficient than shallow ones for certain types of problems.

2012

The potential of deep learning was showcased on a global stage when AlexNet, a deep convolutional neural network (CNN), won the ImageNet competition by a significant margin. The exceptional performance of AlexNet underscored the unparalleled capabilities of deep learning models in image classification tasks. This watershed moment not only bolstered faith in CNNs but also catalyzed a surge in AI research and innovations, solidifying the position of deep learning at the forefront of modern AI advancements.

2014

Artificial intelligence witnessed a groundbreaking advancement with the introduction of Generative Adversarial Networks (GANs) by Ian Goodfellow, Yoshua Bengio, and collaborators. We've discussed these several times in this book; GANs have become a cornerstone in the field, enabling a myriad of applications ranging from art creation to data augmentation. Another remarkable generative model was also introduced in 2014: Variational Autoencoders (VAEs) were described by Diederik Kingma and Max Welling in their paper "Auto-Encoding Variational Bayes." VAEs reframed the generative modeling process and have since opened the door to a variety of applications, ranging from intricate image synthesis to innovative molecule design. As GANs and VAEs emerged concurrently, the two techniques significantly broadened the horizons of what was achievable in the domain of synthetic data generation as well.

2015

DeepMind's AlphaGo achieved what had been deemed impossible when it became the first computer program to defeat a world champion in the strategically intricate board game of Go. As described in the 2016 paper "Mastering the game of Go with deep neural networks and tree search," by David Silver et al., AlphaGo showcased its prowess by defeating Lee Sedol, one of the game's most decorated players, in four out of five games—a milestone previously believed to be decades away. The innovation did not stop there. In 2017, DeepMind revealed AlphaGo Zero, a superior iteration. Unlike its predecessor, which learned from a combination of human expert games and self-play, AlphaGo Zero's learning was exclusively through self-play, devoid of any human expertise, relying solely on the game's rules. Remarkably, it surpassed the version that defeated Lee Sedol in a mere three days of training. This revolutionary approach was further generalized with the introduction of AlphaZero, which in just a few hours mastered not only Go but also chess and shogi, eclipsing the performance of the best AI models in each domain.

2017

The next wave was the emergence of transformer architectures. Pioneering models like BERT from Google's AI team, along with OpenAI's series of GPT models, have ushered in an era of heightened capability in language understanding and generation tasks. BERT's bidirectional training and the self-attention mechanism intrinsic to transformer architectures capture nuanced context with astounding precision, setting new benchmarks across multiple natural language processing tasks. OpenAI's GPT excels in generating humanlike text, showcasing the model's deep grasp of linguistic structures. A salient characteristic of these models' success is scale. The sheer magnitude of data they're trained on, combined with vast model sizes, is a pivotal factor in their unparalleled performance. These advances have collectively redefined the boundaries of what's achievable in the domain of machine understanding and synthesis of language.

2018

Yoshua Bengio, Geoffrey Hinton, and Yann LeCun were awarded the Turing Award for their work on deep learning.

2020

DeepMind introduced AlphaFold, which marked a historic breakthrough in understanding one of biology's long-standing enigmas: protein folding. Proteins, the essential workhorses within cells, derive their function largely from their three-dimensional conformation. Accurately predicting this three-dimensional structure solely based on an amino acid sequence—a task termed "protein folding"—holds profound significance. It provides key insights into understanding diseases and crafting innovative drugs. At the 2020 Critical Assessment of Protein Structure Prediction (CASP) competition, AlphaFold demonstrated unparalleled precision, rivaling results from traditional experimental techniques. The remarkable achievements of AlphaFold promise transformative implications, spanning various domains of biology and medical science. In the same year, Liquid Networks were introduced in a paper in *Nature Intelligence.*

2021

Generative AI witnessed another leap forward with the introduction of the Stable Diffusion model. The paper "High-Resolution Image Synthesis with Latent Diffusion Models" presents this cutting-edge text-to-image synthesis model designed to craft intricate images directly from textual prompts. OpenAI's renowned DALL-E, on the other hand, is based on the transformer architecture, not Stable Diffusion. Introduced in January 2021, DALL-E created visuals from natural language descriptions. Building on its success, OpenAI subsequently unveiled DALL-E 2, an enhanced version designed to produce even more lifelike images at superior resolutions.

2022

DeepMind introduced AlphaCode, an AI-driven generator of computer programs tailored to participate in competitive programming challenges. Underlying its proficiency is a reinforcement learning framework, which trains the model to predict subsequent tokens in a coding sequence based on the present problem definition and the ongoing code. AlphaCode delivered solutions on par with the upper 54% of human participants in programming contests.

2022–23

The AI field sped through a period of remarkable technological advances, deeper integration into various sectors, huge financial investments, growing societal and ethical discussions, and increased global focus on harnessing and regulating the power of the technology. The intensified global investment underscored its strategic value, even as debates about AI's role in creativity, consciousness, and the job market stimulated intellectual and public discourse. OpenAI introduced the GPT-3 and GPT-4 foundational models. This was a true watershed moment in artificial intelligence. These models transformed the landscape of AI accessibility, making sophisticated AI not just a tool exclusively for tech giants and researchers but an everyday utility for everyone. With the advent of ChatGPT, rooted in these advanced models, AI transitioned from specialized labs and applications to virtually everyone's pockets. This democratization signified more than just technological progress; it impacts the way we communicate, seek information, and interact with digital platforms, heralding a new era where AI seamlessly integrates into daily life. Several new startups in the space of foundational models were launched, including our startup LiquidAI, and many new generative AI products were introduced, including Meta's Llama, the highly advanced chatbot Amazon Q, and Alphabet's Gemini.

Appendix 2

The Infrastructure of AI

JUST AS ROADS, BRIDGES, AND UTILITIES ARE FOUNDAtional to urban development and daily life, we need infrastructure to provide the essential components and frameworks that enable AI algorithms to function, scale, and evolve. This encompasses a range of elements, from data to algorithms to hardware accelerators and storage solutions to software platforms and toolchains. This infrastructure not only supports the deployment of AI models but also ensures their efficiency, reliability, and adaptability in diverse real-world scenarios, and includes critical layers that make AI not just operational but optimally effective.

DATA

AI systems learn by analyzing data, and their success largely hinges on access to relevant, high-quality data. The data that is used to train AI systems can be historical records, real-time input, or a mix of both. Moreover, data comes in a multitude of forms, such as text, images, and readings from sensors. Ideally, this data should be labeled, meaning it should have annotations that indicate the correct output or classification for a given input. These tags should denote what the data represents, such as a picture of a specific animal or object.

Labeled data, also known as ground truth, greatly facilitates supervised learning, wherein AI systems learn to predict outcomes or categorize data based on previous examples. However, when labeled data is not available, AI systems can learn through unsupervised learning techniques, which involve discerning patterns and structures from unlabeled data. An especially potent approach is to use multimodal data, which integrates different types of data to provide a more comprehensive view of a given problem. A textual description can be powerful. An image can be useful, too. But text combined with image is better than one or the other, offering richer context that can improve the accuracy of AI predictions.

Text, in particular, can be a powerful augmenting force for any data source. By analyzing text, AI systems can extract additional context, nuances, and insights that may not be evident from other types of data. Whether it's parsing news articles to enhance stock market predictions or interpreting doctor's notes to improve patient care, text data can amplify the capabilities of AI models and enable them to deliver more insightful, robust, and accurate results.

ALGORITHMS

A rich ecosystem of models and algorithms serve as the building blocks of AI systems. These models and algorithms define the way an AI system learns from data and makes predictions or decisions. Each model or algorithm is designed to solve specific types of problems or work with particular kinds of data.

The algorithms range from basic machine learning techniques like linear regression and decision trees—we looked at the random forest decision tree and others in chapter 11—to more complex methods for predictive AI, generative AI, and reinforcement learning.

Selecting an optimal model or algorithm depends on the problem at hand. While convolutional neural networks (CNNs) are well-suited for image recognition tasks, recurrent neural networks (RNNs) are more appropriate for sequential data such as time series or natural language processing. Reinforcement learning algorithms are designed to make decisions in dynamic environments, making them a good fit for tasks like robotics or gameplay.

TRAINING

Once an appropriate model is chosen, it must be trained on data. Training data is fed into the model, and the model adjusts its internal parameters to learn the patterns within the data. The quality of the training data is crucial. Biased or incomplete data will lead to poor or biased model performance. Training requires extensive computation to process vast datasets and intricate algorithms. As models grow in complexity and data volume increases, advanced computational resources become indispensable.

TESTING

After training, the model is tested on previously unseen data to assess its performance. Evaluation metrics like accuracy, precision, recall, or F1 score help determine how well the model generalizes to new data—how well it performs in situations involving data it has never encountered before. Once the model performs satisfactorily, it can be deployed for real-world applications.

TUNING

In addition to training, the model can be fine-tuned to the specific problem or domain. This involves adjusting hyperparameters, which are parameters that are not learned from the data but are established before the learning process begins. This kind of tuning can have a significant impact on a model's performance, so finding the right settings is an essential part of the AI workflow. And it highlights how we humans continue to be critical!

SOFTWARE

The development and deployment of AI systems require robust software infrastructure, including programming languages, frameworks, libraries, and tools that facilitate the design, implementation, testing, and maintenance of models and algorithms. Open-source software (OSS) plays a crucial role in providing these building blocks. OSS provides a rich collection of languages, frameworks, libraries, and tools that accelerate AI development and deployment, while also promoting collaboration and knowledge-sharing, which has been essential for the rapid progress in AI research and technology. Programming languages such as Python and Julia are popular choices for AI development due to their simplicity, versatility, and powerful data manipulation capabilities. Frameworks like TensorFlow and PyTorch are OSS platforms that offer pre-built components and modules for creating, training, and evaluating machine learning and deep learning models. These frameworks significantly reduce the time and effort required to implement complex AI systems, enabling researchers and developers to focus on experimenting with novel models and algorithms.

AI development is further facilitated by OSS libraries such as NumPy, SciPy, and Pandas for numerical computing, data analysis, and data manipulation. OpenCV is a versatile library for computer vision tasks, and Matplotlib and Seaborn are widely used for data visualization.

COMPUTATION

AI systems require robust computational infrastructure to function effectively. The demands of training and deploying AI models necessitate specialized hardware, scalable storage, and efficient networking. At the heart of AI computation are graphics processing units (GPUs) or custom-designed chips like tensor processing units (TPUs) that are optimized for the parallel processing of vast amounts of data, making them well-suited for the intensive computations of AI tasks. Moreover, powerful central processing units (CPUs) complement GPUs by handling other crucial tasks such as data pre-processing and managing the overall system.

STORAGE

Scalable storage solutions are vital, as AI systems need to store and access large volumes of data. This can be achieved through a combination of on-premises storage and cloud-based services, which allow for efficient data retrieval and offer the advantage of scaling storage capacity as needed.

CONNECTIVITY

Effective networking capabilities ensure that data can be transmitted swiftly between different parts of the system. High-speed Internet connectivity, optimized data transfer protocols, and edge computing—wherein processing occurs closer to the data source—all contribute to minimizing latency and accelerating the AI workflow.

FLEXIBILITY

Finally, AI is an ever-evolving landscape. As a result, AI infrastructure must be flexible and adaptable, capable of accommodating emerging technologies and methods. As new solutions emerge, we need to be able to incorporate them efficiently. With the right combination of hardware, software, and networking, AI systems can learn from data more effectively, leading to more powerful models and a more positive overall impact.

Further Reading

THE FOLLOWING ARE FOUNDATIONAL TEXTS FOR ANYONE interested in the technical nuances of AI and machine learning. I highly recommend them.

Russell, S. J., & P. Norvig, "Artificial Intelligence: A Modern Approach." Prentice Hall, 2009.

Bishop, C. M. *Pattern Recognition and Machine Learning.* Springer, 2006.

Foster, D. *Generative Deep Learning: Teaching Machines to Paint, Write, Compose, and Play.* O'Reilly Media, 2019.

Géron, A. *Hands-On Machine Learning with Scikit-Learn, Keras, and TensorFlow.* O'Reilly Media, Inc., 2019.

Goodfellow, I., Y. Bengio, and A. Courville. *Deep Learning.* MIT Press, 2016.

Mitchell, T. M. *Machine Learning.* McGraw Hill, 1997.

Nielsen, M. A. *Neural Networks and Deep Learning.* Determination Press, 2015.

Raschka, S., and V. Mirjalili. *Python Machine Learning.* Packt Publishing, 2019.

Sutton, R. S., and A. G. Barto. *Reinforcement Learning: An Introduction.* MIT Press, 2018.

Notes

INTRODUCTION

1 **Four billion people:** GSMA, "The Mobile Economy 2023," 2023.

1. SPEED

11 **how generative AI impacts writing:** Shakked Noy and Whitney Zhang, "Experimental evidence on the productivity effects of generative artificial intelligence," *Science* 381, no. 6654 (July 2023): 187–92.

14 **In the insurance industry, one interesting startup:** Gaya (https://www.gaya.ai/) is a startup that provides AI Copilot tools for the insurance industry. This tool is similar to the Github Copilot, but fine-tuned for the insurance industry.

15 **When GitHub . . . surveyed its users:** Eirini Kalliamvakou, "Research: Quantifying GitHub Copilot's impact on developer productivity and happiness," GitHub blog, September 7, 2022.

16 **turned to AlphaFold:** Feng Ren et al., "Alphafold accelerates artificial intelligence powered drug discovery: Efficient discovery of a novel CDK20 small molecule inhibitor," *Chemical Science* 14, no. 6 (January 2023): 1443–52.

17 **settled on a single compound:** Kevin Jiang, "U of T researchers used AI to discover a potential new cancer drug—in less than a month," *Toronto Star*, January 19, 2023.

20 **The delivery service UPS employs:** Matt Leonard, "UPS adds dynamic routing to ORION, saving 2-4 miles per driver," *Supply Chain Dive,* June 22, 2021.

20 **Hamsa Balakrishnan's group:** "US Air Force Pilots Get an Artificial Intelligence Assist with Scheduling Aircrews," *MIT News, July 8, 2021.*

2. KNOWLEDGE

23 **The number of scientific papers:** Dashun Wang and Albert-László

Barabási, "Big Science," in *The Science of Science* (Cambridge: Cambridge University Press, 2021), 4–5.

24 **In robotics and AI we have gone:** "Growth in AI and robotics research accelerates," *Nature Index* 610, no. S9 (October 2022).

26 **At MIT, Pattie Maes and David Karger:** MIT Schwartzman College of Computing, "Artificial intelligence for augmentation and productivity," MIT CSAIL, August 18, 2023.

29 **spent 4,000 hours:** "What it takes to build an AI legal assistant lawyers can rely on," Casetext, May 12, 2023.

31 **determine with 99.35% accuracy:** Sharada P. Mohanty, David P. Hughes, and Marcel Salathé, "Using Deep Learning for Image-Based Plant Disease Detection," *Frontiers in Plant Science* 7 (September 2016): 1419.

31 **apps with such capabilities:** Agrio is one app that identifies plant diseases and pests. Accessed November 30, 2023.

32 **cows tend to cluster:** Mac Schwager et al., "Data-driven identification of group dynamics for motion prediction and control," *Journal of Field Robotics* 25, no. 67 (May 2008): 205–24.

35 **Oracle uses AI:** "Oracle Cloud ERP Builds on Analytics, Automation, and AI Lead: Long Term Investment and Customer-Led Innovations Drive Roadmap," Constellation Research, May 7, 2021.

36 **startup company ClimateAI:** https://www.cnn.com/2023/11/26/tech/ai-climate-solutions/index.html.

37 **Steve Jobs credited the elegant fonts:** " 'You've got to find what you love,' Jobs says," *Stanford News*, June 12, 2005.

3. INSIGHT

40 **AI Physicist:** Tailin Wu and Max Tegmark, "Toward an AI Physicist for Unsupervised Learning," September 2018, arXiv.1810.10525.

43 **Gen Z crowd:** "How Brandwatch is Building Game-Changing New AI Features Powered by GPT," Brandwatch, April 20, 2023.

44 **Emmanuel Mignot:** Emmanuel Mignot, videoconference interview by Gregory Mone, July 18, 2023.

4. CREATIVITY

54 **The philosopher Joanna Zylinska:** Joanna Zylinska, "Art in the age of artificial intelligence," *Science* 381, no. 6654 (July 2023): 139–40.

54 **the singer Tony Bennett:** Tony Bennett interviewed by Terry Gross, *Fresh Air*, NPR, July 26, 2023.

54 **The Adobe designer:** Aaron Hertzmann, "Can Computers Create Art?," CSAIL guest lecture, May 9, 2023.

55 **designing robots:** Tsun-Hsuan Wang, Juntian Zheng, Pingchuan Ma, Yilun Du, Byungchul Kim, Andrew Spielberg, Joshua Tenenbaum, Chuang Gan, and Daniela Rus. "DiffuseBot: Breeding Soft Robots with Physics-Augmented Generative Diffusion Models," in *Proceedings of Neural Information Processing Systems Conference* (NeurIPS), 2023.

58 **Lupe Fiasco:** Aaron Wade, "How it's Made: TextFX is a suite of AI tools made in collaboration with Lupe Fiasco," *Google for Developers*, August 2, 2023.

60 **2023 paper:** Ziv Epstein, Aaron Hertzmann, Investigators of Human Creativity, Memo Akten, Hany Farid, Jessica Fjeld, Morgan R. Frank, et al. "Art and the Science of Generative AI." *Science* 380, no. 6650 (2023): 1110–11.

60 **"You can do more with these models":** Steve McDonald, videoconference interview by Gregory Mone, July 3, 2023.

60 **60% of today's jobs:** David Autor, "The Labor Market Impacts of Technological Change: From Unbridled Enthusiasm to Qualified Optimism to Vast Uncertainty," *SSRN Electronic Journal*, May 2022.

61 **A study on creative product design work:** François Candelon, Lisa Krayer, Saran Rajendran, and David Zuluaga Martínez, "How People Can Create—and Destroy—Value with Generative AI," Boston Consulting Group, September 21, 2023.

5. FORESIGHT

64 **something similar for self-driving cars:** Alexander Amini et al., "Deep evidential regression," *Proceedings of the 34th International Conference on Neural Information Processing Systems*, December 2020, 14927–37.

65 **predict the properties of new molecules:** Ava P. Soleimany et al., "Evidential Deep Learning for Guided Molecular Property Prediction and Discovery," *ACS Central Science* 7, no. 8 (July 2021): 1356–67.

67 **AI could be the secret to a new approach:** Songtao He et al., "Inferring high-resolution traffic accident risk maps based on satellite imagery and GPS trajectories," *Proceedings of the 2021 IEEE/CVF International Conference on Computer Vision*, 11977–85.

68 **the long-term impacts of genetic variations:** Carles A. Boix et al., "Regulatory genomic circuitry of human disease loci by integrative epigenomics," *Nature* 590 (February 2021): 300–07.

68 **offer some foresight:** Eyal Simonovsky et al., "Predicting molecular mechanisms of hereditary diseases by using their tissue-selective manifestation," *Molecular Systems Biology* 19, no. 8 (August 2023): e11407.

68 **2018 Medium essay:** Michael Jordan, "Artificial Intelligence—The Revolution Hasn't Happened Yet," *Medium*, April 19, 2018.

69 **provide local flood forecasting:** E. Basha and D. Rus, "Design of early warning flood detection systems for developing countries," *Proceedings of the 2007 International Conference on Information and Communication Technologies and Development*, Bangalore, India, 2007, 1–10, doi: 10.1109/ICTD.2007.4937387.

70 **the oceanographer Raymond Schmitt:** Laifang Li, Raymond W. Schmitt, and Caroline C. Ummenhofer, "Skillful Long-Lead Prediction of Summertime Heavy Rainfall in the US Midwest From Sea Surface Salinity," *Geophysical Research Letters* 49, no. 13 (July 2022): e2022GL098554.

71 **In 2016, Dr. Schmitt and his colleagues:** Laifang Li et al., "North Atlantic salinity as a predictor of Sahel rainfall," *Science Advances* 2, no. 5 (May 2016): e1501588.

72 **outperformed the standard forecasting methods by 92%:** Li, Schmitt, and Ummenhofer, "Skillful Long-Lead Prediction."

72 **Bill & Melinda Gates Foundation grant:** "Salient Receives Grant to Provide Enhanced Weather Forecasts for Smallholder Farmers in East Africa," *Cision PR Newswire*, October 10, 2023.

73 **Dina Katabi:** Emily Mullin, "AI can spot signs of Alzheimer's before your family does," *MIT Technology Review*, March 19, 2018.

73 **Regina Barzilay:** Steven Zeitchik, "Is artificial intelligence about to transform the mammogram?" *Washington Post*, December 21, 2021.

73 **BlueDot . . . used AI:** Eric Niiler, "An AI Epidemiologist Sent the First Warnings of the Wuhan Virus," *Wired*, January 25, 2020.

73 **Oscar Health uses AI:** Oscar Health, "Our experience applying language models in healthcare has led us to several key learnings (and a growing list of use cases)," *Oscar Continuous Hackathon* (blog), October 26, 2023.

6. MASTERY

75 **badminton technique:** M. Seong, G. Kim, D. Yeo, Y. Kang, H. Yang, J. DelPreto, W. Matusik, D. Rus, S. Kim, "MultiSenseBadminton: Wearable Sensor–Based Biomechanical Dataset for Evaluation of Badminton Performance." *Scientific Data* (under review).

76 **Surgeons can receive AI feedback:** Dani Kiyasseh et al., "A multi-institutional study using artificial intelligence to provide reliable and fair feedback to surgeons," *Communications Medicine* 3, no. 1 (March 2023): 42.

77 **60% of college students:** "Study: 30% of College Students Have Used ChatGPT for Essays," *Government Technology*, January 25, 2023.

77 **44% of teenagers:** "Back to School Survey: 44% of Teens 'Likely' to Use AI to Do Their Schoolwork for Them This School Year," PR Newswire, Junior Achievement USA, July 26, 2023.

77 **GPT-4 scored 1410:** Jackie Snow, "As School Starts, Teachers Struggle With New Kid in the Classroom: ChatGPT," *Messenger*, August 17, 2023.

78 **Kelly Gibson began experimenting:** Pia Ceres, "ChatGPT Is Coming For Classrooms. Don't Panic," *Wired*, January 26, 2023.

78 **a study with GitHub Copilot:** Paul Denny, Viraj Kumar, and Nasser Giacaman, "Conversing with Copilot: Exploring prompt engineering for solving cs1 problems using natural language," *Proceedings of the 54th ACM Technical Symposium on Computer Science Education V.1*, March 2023, 1136–42.

79 **working alongside an AI was more effective:** Majeed Kazemitabaar et al., "Studying the effect of AI Code Generators on Supporting Novice Learners in Introductory Programming," *Proceedings of the CHI Conference on Human Factors in Computing Systems*, April 2023, 1–23.

82 **the AI helped shrink the gap:** Sherry Ruan et al., "Reinforcement learning tutor better supported lower performers in a math task," April 2023, arXiv.2304.04933.

82 **the kinds of conversations they'd have in the real world:** "Customer Stories: Duolingo," OpenAI, March 14, 2023.

82 **a virtual projection:** "Matrix Holograms: The tutors using Artificial Intelligence to teach," YouTube video streamed by TBD Media Group, January 17, 2023, at 9:33.

85 **virtual coaches for sports:** Thomas Schack, J. E. H. Junior, and Kai Essig, "Coaching with virtual reality, intelligent glasses and neurofeedback: The potential impact of new technologies," *International Journal of Sport Psychology* 51, no. 6 (December 2020): 667–88.

7. EMPATHY

88 **a study of 5,179 customer support agents:** Erik Brynjolfsson, Danielle Li, and Lindsey R. Raymond, "Generative AI at work," *National Bureau of Economic Research*, no. w31161 (October 2023).

89 **the patients preferred:** John W Ayers et al., "Comparing physician and artificial intelligence chatbot responses to patient questions posted to a public social media forum," *JAMA Internal Medicine* 183, no. 6 (June 2023): 589–96.

89 **communicate with patients in a more compassionate way:** Gina Kolata, "When Doctors Use a Chatbot to Improve Their Bedside Manner," *New York Times*, June 13, 2023; Peter Lee et al., *The AI Revolution in Medicine and Beyond* (Indianapolis: Pearson, 2023).

90 **D. Fox Harrell demonstrated the possibilities:** Randy Kennedy, "Meeting 'the Other' Face to Face," *New York Times*, October 26, 2016.

91 **an interactive narrative:** Caglar Yildirim et al., "Toward Computationally-Supported Roleplaying for Perspective-Taking," *Pro-*

ceedings of the International Conference on Human-Computer Interaction, July 2023, 154–71.

91 **AI emotion recognition:** Matthew Groh et al., "Computational Empathy Counteracts the Negative Effects of Anger on Creative Problem Solving," *Proceedings of the 10th International Conference on Affective Computing and Intelligent Interaction*, August 2022, 1–8.

93 **the language of sperm whales:** Jacob Andreas et al., "Toward understanding the communication in sperm whales," *Iscience* 25, no. 6 (June 2022): 104393.

94 **research library of all underwater sounds:** Miles J. G. Parsons et al., "Sounding the call for a global library of underwater biological sounds," *Frontiers in Ecology and Evolution* 10 (February 2022): 39.

9. GENERATING

116 **ten thousand dimensions:** Tom Brown, Benjamin Mann, Nick Ryder, Melanie Subbiah, Jared D. Kaplan, Prafulla Dhariwal, Arvind Neelakantan, et al., "Language models are few-shot learners," *Advances in Neural Information Processing Systems* 33 (2020): 1877–1901.

117 **a breakthrough idea:** Ashish Vaswani et al., "Attention is all you need," in *Advances in Neural Information Processing Systems* 30 (June 2017): 5998–6008.

125 **the Pope in a white puffer coat:** Drake Bennett, "AI Deep Fake of the Pope's Puffy Coat Shows the Power of the Human Mind," *Bloomberg News*, April 6, 2023.

10. OPTIMIZING

129 **teach themselves through reinforcement learning instead:** Wilko Schwarting et al., "Deep latent competition: Learning to race using visual control policies in latent space," February 2021, arXiv:2102.09812.

130 **250 years' worth of games:** "OpenAI Five defeats Dota 2 world champions," OpenAI, April 15, 2019.

131 **analyze radiology results:** Adam Yala et al., "Optimizing risk-based breast cancer screening policies with reinforcement learning," *Nature Medicine* 28, no. 1 (January 2022): 136–43.

131 **optimize strategies for trading stocks:** Manuela Veloso et al., "Artificial intelligence research in finance: discussion and examples," *Oxford Review of Economic Policy* 37, no. 3 (September 2021): 564–84.

131 **40% reduction in cooling costs:** Richard Evans and Jim Gao, "DeepMind AI Reduces Google Data Centre Cooling Bill by 40%," Google DeepMind, July 20, 2016.

11. DECIDING

137 **Random Forest algorithm:** Leo Breiman, "Random forests," *Machine Learning* 45, (October 2001): 5–32.

139 **between 100 million and 200 million:** Larry Greenemeier, "20 Years after Deep Blue," *Scientific American*, June 2, 2017.

A BUSINESS INTERLUDE: THE AI IMPLEMENTATION PLAYBOOK

141 **Salman Khan:** The Indian Deepfaker, "Deepfakes: The Future of Brand Marketing," *Medium*, September 10, 2022.

141 **Charles Barkley:** "Think Like a Player Takes on New Meaning as FanDuel Partners with Charles Barkley for NBA Playoff Ad Campaign," PR Newswire, FanDuel Group, April 10, 2023.

142 **the doctors and the AI system achieved:** "Better Together," Harvard Medical School, News and Research, June 22, 2016.

142 **tools that automate data entry:** Gaya, AI Copilot for Insurance.

12. THE DARK SIDE OF SUPERPOWERS

163 **Hany Farid:** Gregory Mone, "Outsmarting Deepfake Video," *Communications of the ACM* 66, no. 7 (2023): 18–19.

164 **40,000 possible toxic agents:** Fabio Urbina et al., "Dual use of artificial-intelligence-powered drug discovery," *Nature Machine Intelligence* 4, no. 3 (March 2022): 189–91.

165 **launch attacks with unprecedented speed:** "OPWNAI: Cybercriminals starting to use ChatGPT," Check Point Research, January 6, 2023.

165 **concerns about privacy and individual rights:** Dahlia Peterson and Samantha Hoffman, "Geopolitical implications of AI and digital surveillance adoption," Brookings, June 2022; Martin Beraja et al., "Exporting the Surveillance State via Trade in AI," National Bureau of Economic Research, September 2023.

165 **popping up on Amazon:** Seth Kugel and Stephen Hiltner, "A New Frontier for Travel Scammers: AI-Generated Guidebooks," *New York Times*, August 5, 2023.

165 **digital records of a person's Facebook likes could be used:** Michal Kosinski, David Stillwell, and Thore Graepel, "Private traits and attributes are predictable from digital records of human behavior," *Proceedings of the National Academy of Sciences* 110, no. 15 (March 2013): 5802–05.

166 **Cambridge Analytica:** Zoe Kleinman, "Cambridge Analytica: The story so far," *BBC News*, March 21, 2018.

166 **In 2010, a trader deployed:** Silla Brush, Tom Schoenberg, and Suzi Ring, "How a Mystery Trader with an Algorithm May Have Caused the Flash Crash," *Bloomberg News*, April 22, 2015.

167 **significant threats to democracy itself:** Miles Brundage et al., "The Malicious Use of Artificial Intelligence: Forecasting, Prevention, and Mitigation," February 2018, arXiv:1802.07228.

167 **The security leader Trend Micro:** Forward-Looking Threat Research Team, "Codex Exposed: Helping Hackers in Training?," *Trend*, February 3, 2022.

167 **AI-generated emails actually proved more effective:** Lily Hay Newman, "AI Wrote Better Phishing Emails Than Humans in a Recent Test," *Wired*, August 7, 2021.

172 **share some of their findings:** "GPT-4 System Card," OpenAI, March 23, 2023.

173 **the companies' efforts to share:** Daniel Fabian, "Google's AI Red Team: The ethical hackers making AI safer," *Google: The Keyword*, July 19, 2023.

13. TECHNICAL CHALLENGES

178 **labeling is done by lower-wage workers:** Billy Perrigo, "Open AI Used Kenyan Workers on Less than $2 Per Hour to Make ChatGPT Less Toxic," *Time*, January 18, 2023.

179 **the VISTA simulator:** Alexander Amini et al., "Vista 2.0: An open, data-driven simulator for multimodal sensing and policy learning for autonomous vehicles," *Proceedings of the International Conference on Robotics and Automation*, November 2021, 2419–26.

180 **interesting feature:** Tsun-Hsuan Wang, Alaa Maalouf, Wei Xiao, Yutong Ban, Alexander Amini, Guy Rosman, Sertac Karaman, and Daniela Rus, "Drive Anywhere: Generalizable End-to-End Autonomous Driving with Multi-modal Foundation Models," arXiv preprint arXiv:2310.17642 (2023).

181 **626,000 pounds:** Emma Strubell, Ananya Ganesh, and Andrew McCallum, "Energy and Policy Considerations for Deep Learning in NLP," *Proceedings of the 57th Annual Meeting of the Association for Computational Linguistics*, January 2019, 3645–50.

181 **700,00 liters:** Pengfei Li, Jianyi Yang, Mohammad A. Islam, and Shaolei Ren, "Making AI Less Thirsty: Uncovering and Addressing the Secret Water Footprint of AI Models," 2023, arXiv:2304.03271.

182 **MosaicML:** Generative AI for All, Mosaic, accessed October 4, 2023, www.mosaicml.com.

182 **LiquidAI:** Aaron Pressman and Jon Chesto, "Boston's answer to ChatGPT: MIT spinoff Liquid AI has a radical new approach," *Boston Globe*, December 6, 2023; see also www.liquid.ai.

182 **The team at MosaicML:** J. Frankle, G. K. Dziugaite, D. M. Roy, and M.

Carbin, "Pruning Neural Networks at Initialization: Why are We Missing the Mark?," International Conference on Learning Representations, 2021, arXiv:2009.08576.

182 **pruning models:** Cenk Baykal, Lucas Liebenwein, Igor Gilitschenski, Dan Feldman, and Daniela Rus, "Sensitivity-Informed Provable Pruning of Neural Networks," SIAM Journal on Mathematics of Data Science 4, no. 1 (2022): 26–45.

182 **Liquid Networks:** Ramin Hasani et al., "Closed-form continuous-time neural networks," *Nature Machine Intelligence* 4, no. 11 (November 2022): 992–1003

184 **The Liquid Networks even adapted:** Makram Chahine et al., "Robust flight navigation out of distribution with liquid neural networks," *Science Robotics* 8, no. 77 (April 2023): eadc8892.

184 **Data distillation:** Noel Loo et al., "Efficient dataset distillation using random feature approximation," *Advances in Neural Information Processing Systems* 35 (October 2022): 13877–91.

185 **BarrierNet:** Wei Xiao et al., "BarrierNet: A safety-guaranteed layer for neural networks," November 2021 arXiv:2111.11277.

186 **biased against Black individuals:** Julia Angwin et al., "Machine Bias," *ProPublica*, May 23, 2016.

186 **less likely to display high-paying job opportunities:** Amit Datta, Michael Carl Tschantz, and Anupam Datta, "Automated experiments on ad privacy settings: A tale of opacity, choice, and discrimination," August 2014, arXiv:1408.6491.

14. SOCIETAL CHALLENGES

192 **Researchers at Stanford:** "What is Fair Use?," Stanford University Libraries.

192 **Shein, a fast-fashion retailer:** Harri Weber, "Designers sue Shein over AI ripoffs of their work," *Y! News: TechCrunch*, July 14, 2023, https://www.npr.org/2023/07/15/1187852963/shein-rico-racketeering-lawsuit.

193 **In a related study:** Aparna Balagopalan et al., "Judging facts, judging norms: Training machine learning models to judge humans requires a modified approach to labeling data," *Science Advances* 9, no. 19 (May 2023): eabq0701.

195 **The European Union has already defined:** European Commission, "Regulatory framework proposal on artificial intelligence," Shaping Europe's Digital Future, https://digital-strategy.ec.europa.eu/en/policies/regulatory-framework-ai, accessed October 5, 2023.

199 **500 ml of water:** Pengfei Li, Jianyi Yang, Mohammad A. Islam, and Sha-

olei Ren, "Making AI Less Thirsty: Uncovering and Addressing the Secret Water Footprint of AI Models," 2023, arXiv:2304.03271.

199 **a 2023 study projected:** Alex de Vries, "The growing energy footprint of artificial intelligence," *Joule* 7, no. 10 (October 18, 2023): 219194.

200 **misleading claims about Covid-19 vaccines:** Shannon Bond, "Just 12 People Are Behind Most Vaccine Hoaxes on Social Media, Research Shows," *All Things Considered*, NPR, May 14, 2021.

201 **digital watermarking:** Hadi Salman et al., "Raising the cost of malicious AI-powered image editing," February 2023, arXiv:2302.06588.

15. WILL AI STEAL YOUR JOB?

205 **Amara's Law:** https://www.pcmag.com/encyclopedia/term/amaras-law#; last accessed December 11, 2023.

205 **not be a one-to-one relationship:** Erik Brynjolfsson and Tom Mitchell, "What can machine learning do? Workforce implications," *Science* 358, no. 6370 (December 2017): 1530–34.

205 **30–40% more productive:** "Large Language Models and the End of Programming," YouTube video produced by ACM Chicago, at 1:04:21.

206 ***Science* study on productivity:** Shakked Noy and Whitney Zhang, "Experimental evidence on the productivity effects of generative artificial intelligence," *Science* 381, no. 6654 (July 2023): 187–92.

207 **A working group at Global Partnerships in AI:** "Broad adoption of AI by SMEs," Global Partnership on Artificial Intelligence, 2021.

211 **Goldman Sachs has projected:** Bryce Elder, "Surrender your desk job to the AI productivity miracle, says Goldman Sachs," *Financial Times*, March 27, 2023.

212 **McKinsey predicted:** "The economic potential of generative AI: The next productivity frontier," McKinsey Digital, June 14, 2023.

212 **the job of baking bread:** Maja S. Svanberg, Wensu Li, Martin Fleming, Brian C. Goehring, and Neil C. Thompson, "Beyond AI Exposure: Which Tasks are Cost-Effective to Automate with Computer Vision?," working paper, available at SSRN 4700751 (2024).

212 **BakeBot:** Mario Bollini et al., "Interpreting and executing recipes with a cooking robot," in *Experimental Robotics: The 13th International Symposium on Experimental Robotics* (Cham, Switzerland: Springer International, 2013), 481–95.

214 **60% of workers today . . . didn't exist in 1940:** "New Frontiers: The Origins and Content of New Work, 1940–2018," National Bureau of Economic Research, August 2022.

16. WHAT NOW?

216 **historically underestimated:** Dan Hendrycks, Mantas Mazeika, and Thomas Woodside, "An Overview of Catastrophic AI Risks," June 2023, arXiv:2306.12001.

216 **a superintelligent AI is a possibility:** Yoshua Bengio, "FAQ on Catastrophic AI Risks," blog post, June 24, 2023.

219 **Stanford AI Index:** "The AI Index Report," Stanford University, 2023.

219 **The SERC stories:** MIT Schwarzman College of Computing, "Case Studies in Social and Ethical Responsibilities of Computing."

221 **European Union's General Data Protection Regulation:** See https://gdpr.eu/fines/, accessed December 11, 2023.

222 **measure a model's potential for extreme:** Toby Shevlane et al., "Model evaluation for extreme risks," May 2023, arXiv:2305.15324.

222 **Vint Cerf:** Pallab Ghosh, "Google's Vint Cerf warns of 'digital Dark Age,'" *BBC News*, February 13, 2015.

223 **Microsoft has published a set of guidelines:** "Introduction to red teaming large language models (LLM)," Microsoft, July 18, 2023.

223 **Multiple companies have pledged:** "Fact sheet: Biden–Harris Administration Secures Voluntary Commitments from Eight Additional Artificial Intelligence Companies to Manage the Risks Posed by AI," White House press release, September 12, 2023.

224 **only 2% covered safety:** Zachary Arnold, Jennifer Melot, Dewey Murdick, and Brian Love, "ETO Research Almanac," Center for Security and Emerging Technology, Georgetown University, May 19, 2023.

224 **provide developers with dissent mechanisms:** Roel Dobbe, Thomas Krendl Gilbert, and Yonatan Mintz, "Hard choices in artificial intelligence," *Artificial Intelligence* 300 (November 2021): 103555.

224 **dubbed MACHIAVELLI:** Alexander Pan et al., "Do the Rewards Justify the Means? Measuring Trade-offs Between Rewards and Ethical Behavior in the Machiavelli Benchmark," *Proceedings of the International Conference on Machine Learning*, April 2023, 26837–67.

225 **AI Act:** "Artificial intelligence act: Council and Parliament strike a deal on the first rules for AI in the world," Council of the EU press release, December 9, 2023.

APPENDIX 1: A BRIEF HISTORY OF ARTIFICIAL INTELLIGENCE

241 **Turing test:** Alan M. Turing, "Computing Machinery and Intelligence," *Mind* 59, no. 236 (October 1950): 433–60.

242 **Dartmouth workshop:** "A Proposal for the Dartmouth Summer Research Project on Artificial Intelligence," Stanford University, August 21, 1955.

244 **Neocognitron:** Kunihiko Fukushima, "Neocognitron: A Self-organizing Neural Network Model for a Mechanism of Pattern Recognition Unaffected by Shift in Position," *Biological Cybernetics* 36, no. 4 (April 1980): 193–202.

245 **Hopfield:** J. J. Hopfield, "Neural networks and physical systems with emergent collective computational abilities," *Proceedings of the National Academy of Sciences* 79, no. 8 (April 1982): 2554–58.

246 **backpropagation algorithm:** David E Rumelhart, Geoffrey E. Hinton, and Ronald J. Williams, "Learning representations by back-propagating errors," *Nature* 323, no. 6088 (October 1986): 533–36.

246 **Richard Sutton's:** Richard S. Sutton, "Learning to Predict by the Methods of Temporal Differences," *Machine Learning* 3 (February 1988): 9–44.

247 **LeNet architecture:** Yann LeCun et al., "Gradient-based learning applied to document recognition," *Proceedings of the IEEE* 86, no. 11 (November 1998): 2278–324.

248 **Generative Adversarial Networks:** Ian Goodfellow et al., "Generative Adversarial Networks," June 2014, arXiv:1406.2661.

248 **VAEs:** Diederik P. Kingma and Max Welling, "Auto-encoding Variational Bayes," December 2013, arXiv:1312.6114.

248 **AlphaGo:** David Silver et al., "Mastering the game of Go with deep neural networks and tree search," *Nature* 529, no. 7587 (January 2016): 484–89.

249 **transformer architectures:** Ashish Vaswani et al., "Attention is all you need," *Advances in Neural Information Processing Systems* 30 (June 2017), arXiv:1706.03762.

250 **Liquid Networks:** Mathias Lechner et al., "Neural circuit policies enabling auditable autonomy," *Nature Machine Intelligence* 2, no. 10 (November 2020): 642–52.

250 **Stable Diffusion:** Robin Rombach et al., "High-Resolution Image Synthesis with Latent Diffusion Models," April 2022, arXiv:2112.10752.

251 **GPT-3 and GPT-4:** Sébastien Bubeck et al., "Sparks of artificial general intelligence: Early experiments with GPT-4," April 2023, arXiv:2303.12712.

Index

INDEX